THE GREENHOUSE EFFECT

This book outlines in a clear and objective fashion the reasons behind our strange weather and changes in climate. The evidence linking this with a warming of the planet due to the greenhouse effect, and the likely impacts on Britain and the rest of the world are examined.

A range of 'greenhouse gases' are rapidly building up in the atmosphere. Acting like a thermostat, they are turning up the heat to levels not seen for over 100,000 years. This will bring rising sea levels, increased frequency of storms, floods and droughts, and disruption to agriculture and ecosystems. The world as we know it is changing and at a rate faster than it can adapt to – *unless we act now*.

THE GREENHOUSE EFFECT – A Practical Guide to Our Changing Climate is a guide book to the changes we are inflicting on ourselves and our planet. It is a comprehensive guide and should be essential reading for anyone concerned about the environment or simply curious about a phenomenon likely to dominate our lives in the future.

About the Authors

Stewart Boyle was born in Hawick, Scotland. He became actively involved in environmental issues in 1979 when he set up a local Friends of the Earth group in Coventry. He joined the national staff of Friends of the Earth in 1984 as Energy Campaigner. In this role he was actively involved in defeating the government's nuclear waste dumping plans. He joined the Association for the Conservation of Energy in 1988 where he works on the global warming issue and in promoting energy efficiency measures in order to reduce environmental destruction.

John Ardill was born in Thirsk, Yorkshire and educated locally and in Iraq, Lebanon, Cyprus and Ireland. He has worked for *The Guardian* since 1965 as, successively, regional affairs, labour and environment correspondent. He has written a number of books on the environment.

The Greenhouse Effect

A Practical Guide to Our Changing Climate

Stewart Boyle and John Ardill

Foreword by David Bellamy
Preface by Mostafa K. Tolba
Illustrations by John Barker

NEW ENGLISH LIBRARY
Hodder and Stoughton

Copyright 1989 by Stewart Boyle
and John Ardill

First published in Great Britain in
1989 by New English Library
paperbacks

An NEL paperback original

Second impression 1989

This book is sold subject to the
condition that it shall not, by way
of trade or otherwise, be lent,
re-sold, hired out or otherwise
circulated without the publisher's
prior consent in any form of
binding or cover other than that in
which it is published and without
a similar condition including this
condition being imposed on the
subsequent purchaser.

No part of this publication may
be reproduced or transmitted in
any form or by any means,
electronically or mechanically,
including photocopying, recording
or any information storage or
retrieval system, without either the
prior permission in writing from
the publisher or a licence,
permitting restricted copying. In
the United Kingdom such licences
are issued by the Copyright
Licensing Agency, 33–34 Alfred
Place, London WC1E 7DP.

British Library C.I.P.

Boyle, Stewart
    The greenhouse effect.
    1. Climate. Effects of carbon
    dioxide
    I. Title   II. Ardill, John
    551.6

    ISBN 0-450-50638-X

Printed and bound in Great
Britain for Hodder and Stoughton
paperbacks, a division of Hodder
and Stoughton Ltd., Mill Road,
Dunton Green, Sevenoaks, Kent
TN13 2YA (Editorial Office: 47
Bedford Square, London WC1B
3DP) by Richard Clay Ltd.,
Bungay, Suffolk. Photoset by
Rowland Phototypesetting Ltd.,
Bury St Edmunds, Suffolk.

This book is dedicated to the children who inherit the world we have spoiled, with the hope that they will learn to make it better.

CONTENTS

ACKNOWLEDGEMENTS

Writing this book has necessarily depended on the assistance and patience of a wide range of people. Our only hope is that they feel the attempt to explain such a complex issue to a wider audience has justified their sacrifices. We particularly thank Bernadette Vallely and Elizabeth Ardill for unstinting support and encouragement throughout; Chris Rose for giving Stewart the confidence to take on the project in the first place; John Barker for his excellent graphics; Debbie Jellings and the Women's Environmental Network office for typing and logistical support; Mike Harper for additional research and the appendices; Bob Watson, Jose Lutzenberger, George Woodwell and James Lovelock for interviews despite busy schedules; to everyone at ACE for support and understanding; to Mike Flood, Koy Thompson, Fiona Weir, Ian Fairlie, Andrew Warren, Linda Taylor, Kerry Chester, Jackie Karas and Mary Blake for comments on earlier drafts; to Jim Riccio, Laura Morris from the Oak Ridge CO_2 Information Analysis Centre, Annie Roncerel, Bill Chandler, Ted Hollis, Andy Kerr, Francis Balfour, Mike Grubb, Nigel Mortimor, the Greenpeace UK Press Office, Liz Cook, Rafe Pomerance, Norman Myers, Mick Kelly, Michael Oppenheimer, Tom Stoel, Fred Pearce, John Gribben, Tim Radford and Pippa Hyam for information; to the FoE Rainforest Campaign team for both information and solid support; to Toby Moore for being honest enough to say no; to our agent Bill Hamilton; and finally to Roland Philipps, Broo Doherty and Anna Bence-Trower at Hodder and Stoughton (NEL) for enthusiastic support.

FOREWORD

It is January 1989 and I am writing this foreword while filming in New Zealand. Cool, wet unseasonal weather is ruining the holiday trade in Auckland and the sub-tropical north, while droughts along the east coast of South Island are sending the farmers of the Canterbury Plains into the despair of bankruptcy. I telephone home to the Pennines of north-east England to be told that the garden is full of spring flowers. Both the USA and the USSR have had catastrophic harvests; for every tonne of wheat grown by the farmers of Australia, 13 tonnes of soil are destroyed; by the end of this sad century one-third of all the world's farmland will be unproductive semi-desert; 100,000 people died today of conditions relating to malnutrition and environmental pollution.

The profligate demands of humankind are causing far-reaching changes to the environment of planet Earth, of this there is no doubt.

One of the symptoms of this global malaise is that the Earth's temperature is showing an upward swing, the so-called greenhouse effect, now a subject of international concern.

The greenhouse effect may melt the glaciers and ice-caps of the world causing the sea to rise and flood many of our great cities and much of our best farmland. The greenhouse effect may be simply holding off the next ice age, an equally disturbing scenario.

Read all about it so that you can take part in the great debate and be well informed and willing to take the necessary steps to avoid catastrophe.

David J. Bellamy
Massey University, 1989

PREFACE

These last years of the twentieth century are a crucial time both for the natural foundations of the planet and for human understanding of, and responses to, environmental challenges. The depletion of the ozone layer is a harsh warning that our knowledge of the complex planetary systems is still incomplete. Thankfully, protection of the ozone layer now has high priority for many governments, and I am confident that we can revise the Montreal Protocol to incorporate stronger controls of chemicals which damage the layer.

Ozone layer depletion has become a vivid symbol of our ability to upset the workings of our planet. But it has also produced evidence that nations can respond quickly to environmental challenge. The greenhouse effect, and its consequent global warming and climate change, has causes as varied as tropical forest destruction, urbanisation, economic growth, intensive agriculture, and the use of CFCs, also the agents of ozone depletion. Thus no one should have any illusions about the difficulty of containing climate change. It will require little less than a new global ethic: economic growth which does not threaten nature. Hence this timely book goes to the heart of the environmental agenda which now confronts us all.

Dr Mostafa K. Tolba
Executive Director
United Nations Environment Programme
1989

1

THE WEATHER CONQUERS ALL

At 65,000 feet, Concorde flies above the weather, parting the thin cold air of the stratosphere. Its passengers are cocooned in comfort, held aloft by the wonders of modern technology, the accumulated knowledge of 250 years of scientific discovery and industrial development which has dared to master the elements – and is coming close to destroying them.

Somewhere below, three-quarters of Bangladesh is under water, and a quarter of its people are without homes. Three thousand have died in the monsoon floods and, in a few weeks time, 5,000 more will perish as a tropical cyclone brings the Bay of Bengal surging through its mangrove forests and into its southern cities. Some time next century ten per cent of the country could be permanently submerged.

On the Rio Grande, Texas ranchers are blow-torching the thorns off the prickly pear cactus to let their parched and starving cattle get at the soft flesh. Some of the cacti are too dry to bother with. The corn and soy-beans are withered in Dakota. Three thousand barges are grounded in the Mississippi and the Governor of Illinois has raised a political storm by asking for water to be diverted from the Great Lakes. In Alpine township, Michigan, Mexican migrant workers are sweltering in the tropical heat, waiting forlornly for the rain which might revive the fruit and vegetable harvest. Up in Slave Lake, Alberta, they are wallowing in the mud from the flooded

Sawridge Creek. No one can remember anything like it.

In the Anhui Province of China, 70,000 wells have run dry but elsewhere in the country more than 1,400 have died in floods and storms. In the southern city of Xian the rice ration will be one kilogramme a month. The authorities say 20 million are facing possible starvation. When the Chinese start to plough their fields, the dust is visible in Hawaii.

The blue waters of the Caribbean are stained red by the rivers of silt washed off the once wooded slopes of Haiti. In recent years, a million Haitians have fled from the fields that seem to grow stones.

Honeymooners cower in their Jamaican hotels as windows, doors and walls collapse around them. One in four Jamaicans is homeless and the island's economy is in tatters in the wake of Hurricane Gilbert. In the heart of the storm US and Soviet weather planes swap notes and try to keep out of one another's way as they measure the hurricane's brutal strength. Ideologies are redundant in the face of the weather, which makes all men and women equal.

Hurricane Joan has flattened 15,000 square kilometres of forest of Nicaragua. The town of Bluefields is in ruins. The rivers are choked, the land is a brown gruel of mud and smashed vegetation. It looks like Hiroshima after the bomb.

From Yellowstone National Park to Amazonia the skies are filled with the smoke and stench of forest fires. The former are caused by an accident of nature, the latter by the design of man. In twenty years Amazonia may be as bereft of trees as Haiti. The rubber tappers' leader Chico Mendes is shot down for standing, Gandhi-like, in the path of the chainsaws. By the end of the century India too could be treeless.

In the drought-stricken heart of Africa, where in recent years 10 million have sought refuge from famine in camps and shanty towns, 200mm of rain has fallen on Khartoum in less than twenty-four hours. More than a million are homeless. Gabon has had its first floods in twenty years. The rains have dissolved mud houses right across the parched Sahel, but the crops of millet and sorghum are waving green and tall as the surviving rooftops. In Senegal and Mauritania, the Gambia and Guinea Bissau, Niger and Mali, the locusts are breeding. Desert Africa is anticipating the worst locust plague on record.

For the first time in a decade, India is buying grain on the world market. Well ahead of schedule, the Soviet grain buyers are doing the rounds of Canada, the US and Europe. The world's granaries are emptier than they have been for years. Another year's drought in the Northern croplands could push the world into hunger.

Melbourne catches a dying fragment of the annual Antarctic ozone hole. Up in Queensland, two out of three are already marked for sun-induced skin cancers; and the sky's natural filter is being blown away by chemicals used in certain spray cans, soft chairs and air conditioners.

Around the globe, countless chimneys and exhaust pipes, cooking stoves and bonfires are pouring a colourless gas called carbon dioxide into the air. Every lump of coal, every tankful of petrol, every log, every wisp of straw that is put to flame adds to its load. Felled forests and newly cultivated fields add more. Stockyards and sheep farms, rice paddies and rubbish dumps, contribute another gas called methane. Ploughed and fertilised fields add a third, nitrous oxide.

Below Concorde's flightpath, in the troposphere – the atmospheric weather zone – the air is thickening and warming itself with all these gases. Meanwhile, molecules of man-made gases called chlorofluorocarbons (CFCs) used in aerosols and fridges drift upwards to the higher atmosphere to break apart in the sunlight. Each atom of chlorine they release will bounce around the stratosphere smashing 100,000 molecules of ozone.

In London, Prime Minister Margaret Thatcher has

told scientists that we may have 'unwittingly begun a massive experiment with the system of this planet'. She is not the first to make this observation. In Toronto, politicians, scientists and administrators from around the world have called for a halving of carbon dioxide emissions from man-influenced sources. On the campaign trail, George Bush promised a global conference about the problem once he reached the White House. We await developments. The Soviet Foreign Minister Eduard Schevardnadze tells the General Assembly of the United Nations that 'the growing physical destruction of our planet is the verdict against the existing divisions of the world'.

As 1988 turned, Boxing Day swimmers in London's Serpentine complained that the water was too warm. In Alaska, the thermometer sank to a record 50°C below zero. In Italy, the driest winter for seventy years sent ministers in search of some of those declining stocks of cereals. Water was being rationed in Sardinia where it had not rained for fifteen months. Snowless ski resorts faced bankruptcy and on the Cote d'Azur fire-fighting planes were pouring water on burning forests. A high-pressure system was sitting over Northern Italy, Southern Germany and the Alps. In typically understated fashion the British Meteorological Office said, 'I suppose one would call it random chance that it has persisted.'

Many of 1988's droughts and floods, heat waves and hurricanes, were random events, the fall of the dice. But the dice are being weighted. In coming years, they will fall hot and stormy-side uppermost more often. Many of the events were random, but scientists and politicians began to see connections. Hard-nosed politicians with voters to cosset, powerful vested interests to satisfy, and rivals to guard against, began to talk in the language of prophets, ecologists and utopians. They began to talk about a world which cherishes its resources like a miser, and outlaws waste. They began to talk of a world which is frugal and fair. They began to talk about mankind laying down both its weapons of war and words of dissent, to labour as one in tackling the threat it is making against itself.

In the summer of 1988, the heads of all the United Nations agencies met in Oslo to consider the future of the world. As the UN's priorities for the following twelve years they identified: developing human resources and fully integrated population policies; protecting the atmosphere and the global climate, ocean and water resources; halting desertification and countering deforestation; controlling dissemination of dangerous wastes and aiming at the elimination of such wastes; increasing co-operation in the development of technology; controlling soil erosion and the loss of species; and, above all, securing economic growth, social justice and a more equitable distribution of income and resources within and among countries as a means for alleviating poverty.

Secretary-General Javier Perez de Cuellar and the Norwegian Prime Minister Gro Harlem Brundtland said in a joint statement:

> To achieve these goals a new global ethic is needed based on equity, accountability and human solidarity – solidarity with present and future generations – rather than on the tyranny of the immediate.

Noble sentiments are two-a-penny in the mouths of prominent figures. Poverty, inequality and environmental degradation have long been with us. But we believe that in 1988 something lent sincerity and urgency to these pronouncements. In 1988 the atmosphere came within one per cent certainty of proving that humanity has upset its natural balance and that it will strike back blindly and with catastrophic unpredictability. Global warming is the threat which bundles up all our woes into a single problem and a single solution.

Mostafa Tolba, Executive Director of the United
Nations Environment Programme (UNEP) has said:

> A global sense of urgency is developing. The public
> concern over global warming is mounting at an un-
> precedented speed . . . Political leaders now accept the
> broad scientific consensus that human activity is alter-
> ing climate and that the changes and their impacts will
> become more pronounced over the next few decades.
> The warming warning is being heeded.
>   . . . There are about 4,000 days left in this century.
> But the scale of the challenge before us means that we
> shall all need to work, in one way or another, on every
> day of the coming decade if we are to meet these
> challenges.

This book is about climatic change: what is causing it,
what it will do to us, and what we can do about it. It looks
at the behaviour of the atmosphere, the scientific uncertainties, the impacts of climatic change on nature and
society, and the choices that lie before us. It attempts to
trace some of the threads which bind together the living-
rooms of Manhattan and the shanty towns of Rio; the
paddy fields of Japan and the hay meadows of Iceland;
the rainforests of Brazil and the coffee tables of Ken-
sington. It is about the challenges which unite us all, the
way the world is responding, and the course we must
attempt to steer in the years ahead.

2

THE SCIENCE OF THE GREENHOUSE

> Whenever people talk to me about the weather, I always feel
> certain that they mean something else.
>
> Oscar Wilde

People all over the world love talking about the weather, particularly bad weather. As natural disasters and freak weather conditions have piled up unremittingly over the past couple of years, they have provided almost unlimited scope for conversation. When we talk about 'the funny weather we're having', we are usually thinking back to what it was like in previous years. If we are old enough we invariably hark back to the days of our youth when it was, of course, better – or worse – than it is now. Our memories don't necessarily play us false. The weather *does* change from year to year, and over longer periods of time. A long life may well encompass significant alterations to the weather – the winters growing milder, the summers wetter, as the climate slowly changes.

Technically speaking, **weather** is what is happening right now: the current state of the atmosphere which is displayed daily by the TV meteorologists. **Climate** describes the general characteristics of the weather experienced in any region: its variations in temperature, humidity, wind direction and strength, and precipitation. Climate takes account of rare, more extreme events which cannot always be predicted with much accuracy

The Greenhouse Effect

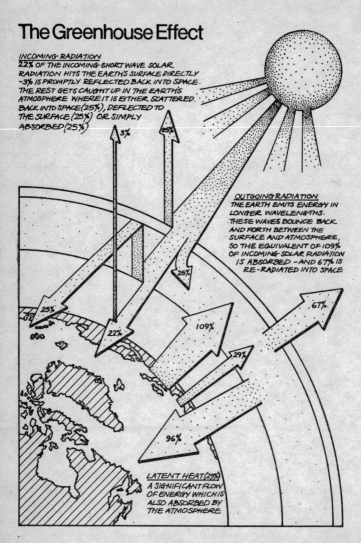

INCOMING RADIATION
22% OF THE INCOMING SHORT WAVE SOLAR
RADIATION HITS THE EARTH'S SURFACE DIRECTLY
-3% IS PROMPTLY REFLECTED BACK INTO SPACE.
THE REST GETS CAUGHT UP IN THE EARTH'S
ATMOSPHERE WHERE IT IS EITHER SCATTERED.
BACK INTO SPACE (25%), DEFLECTED TO
THE SURFACE (25%) OR SIMPLY
ABSORBED (25%)

OUTGOING RADIATION
THE EARTH EMITS ENERGY IN
LONGER WAVELENGTHS.
THESE WAVES BOUNCE BACK
AND FORTH BETWEEN THE
SURFACE AND ATMOSPHERE,
SO THE EQUIVALENT OF 109%
OF INCOMING SOLAR RADIATION
IS ABSORBED - AND 67% IS
RE-RADIATED INTO SPACE

3%

25%

25%

25%

22%

109%

67%

29%

96%

LATENT HEAT (29%)
A SIGNIFICANT FLOW
OF ENERGY WHICH IS
ALSO ABSORBED BY
THE ATMOSPHERE

Fig. 1

but can, on the basis of experience, be given a probability of happening once every so many years.

This book is mainly about climate, and how it is changing due to global warming as a result of a phenomenon called the **greenhouse effect** (see fig. 1). The climate we are accustomed to is crucial to the way we live, to our health, to our ability to grow food, and to the wildlife and landscapes which are the background to our daily lives. It influences the style of our buildings and our cities. It shapes the way we think of ourselves, and of others, and the way we behave. The brisk, phlegmatic Briton, the passionate Spaniard, the dour Finn, are all creatures, at least in part, of their climate and their landscape. Knowledge of our climate determines long-term decisions on issues as varied as the crops we grow, the type of houses we build, the sort of surfaces we put on our roads, the size of our electricity industry, and the amount of water we keep in our reservoirs.

Climatologists like Tom Wigley, head of the Climatic Research Unit at the University of East Anglia, examine decades of actual weather reports, rainfall and temperature records and the evidence of past climatic conditions locked in trees, sediments, ice cores and rocks, to determine long-term trends. They try to isolate the causes behind fluctuations in past climates to tell us whether we're moving towards a warmer, colder, wetter or windier future. Wigley and many of his colleagues around the world are now virtually certain that significant climatic change is on the cards. Those extreme events of 1988 – the droughts, tornadoes and floods – may be part of the change which will give us different climates and different weather.

The Earth has already warmed by more than half a degree centigrade (0.5°C) since the mid-1800's. The reasons behind this rise are complex and will be described shortly, but they lie in our rapid industrial development and use of fossil fuels. A further 1.5 to 4.5°C increase before the middle of the next century is now predicted (see fig. 2). This may not sound much, but it would take the planet to higher temperatures than it has known for more than 100,000 years. A drop in temperature by only 1 °C was enough to cause the **Little Ice Age** in the 1600s and 1700s when the Thames froze over, and the ice was thick enough for horse-drawn carriages to drive across. The last ice age which reached its peak 18,000 years ago was only 4 °C colder than today.

An increase in the global average temperature is only the beginning of the story. Messing about with an atmospheric system which has kept the world comfortable for thousands of millions of years will have more complex consequences. It will not simply be a bit warmer everywhere, all the year round. The temperature increases will be bigger towards the poles than in the tropics, and greater in winter than in summer. Changing the distribution of heat around the surface of the globe will dislocate the weather systems which drive the winds, create storms, and determine the water cycles. A general warming will melt glaciers and expand the volume of water in the oceans. If this happens, the oceans will spill over our low-lying coastal plains and deltas, areas which contain some of the world's biggest concentrations of population and most productive sources of food. It will alter our local climate and perhaps change the shape of our homeland.

While three decades of research have convinced climatologists that the Earth will become substantially warmer, they are much less certain about what will happen in any particular part of the globe. So unpredictable are the effects on even relatively small patches of the Earth that, while it is possible that southern Britain may find itself with a Mediterranean climate sometime during the next century, it is also possible that an alteration in the ocean currents could eventually make the British Isles cooler and wetter than they are now.

What, then, is happening to our atmosphere? What is this greenhouse effect which is expected to bring about

Global Warming

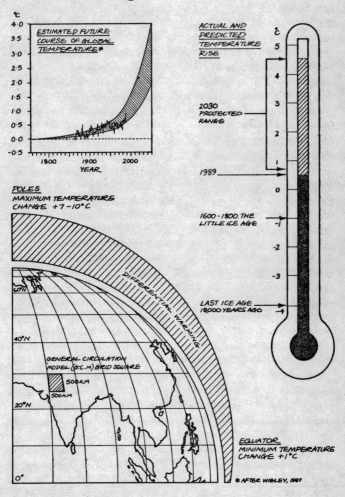

℃

ESTIMATED FUTURE COURSE OF GLOBAL TEMPERATURE*

4.0
3.5
3.0
2.5
2.0
1.5
1.0
0.5
0.0
-0.5

1800 1900 2000

YEAR

ACTUAL AND PREDICTED TEMPERATURE RISE

2030 PROJECTED RANGE

1989

1600 - 1800, THE LITTLE ICE AGE

LAST ICE AGE 18000 YEARS AGO

℃
5
4
3
2
1
0
-1
-2
-3
-4

POLES
MAXIMUM TEMPERATURE CHANGE +7 - 10°C

DIFFERENTIAL WARMING

60°N

40°N

GENERAL CIRCULATION MODEL (G.C.M.) GRID SQUARE

500KM
500KM

20°N

0°

EQUATOR
MINIMUM TEMPERATURE CHANGE +1°C

* AFTER WIGLEY, 1987

Fig. 2

such profound changes? And why can't we be more certain about it?

Stephen Schneider of the National Center for Atmospheric Research in Boulder, Colorado claims that, in spite of all the controversy that surrounds the term, the greenhouse effect is not a scientifically controversial subject. 'In fact, it is one of the best, most well established scientific theories in the atmospheric sciences.' Venus, with a very dense carbon dioxide atmosphere, has ovenlike temperatures at its surface. Mars, with its very thin carbon dioxide atmosphere, has temperatures comparable to our polar winter. The reason why Venus is too hot, Mars too cold, and the Earth *just right* is known to be the result of the greenhouse effect.

The effect works because some gases in the atmosphere allow sunlight through to heat the surface of the planet, but trap the warmth that radiates back into space (see fig. 1). Increase the concentration of these gases, and you increase the amount of heat that is trapped in the lowest part of the atmosphere.

The greenhouse effect was first described by Baron Jean Baptiste Fourier in 1827. The French mathematician pointed out that the atmosphere acted like the transparent glass cover of a box. Experiments and observations during the nineteenth century identified carbon dioxide and water vapour as the gases responsible for this phenomenon, and prompted the first speculations that variations in the atmospheric concentration of these gases might account for periods of altered climate like the ice ages. In 1896 a Swedish scientist, Svante Arrhenius, suggested that industrial pollution might, over a period of centuries, double the amount of carbon dioxide in the atmosphere and raise the global temperature by 5°C. Cold regions like Sweden, he suggested, might hope to enjoy ages with more equable climates. The essence of his theory 'has yet to be disproved,' says Veerabhadran Ramanathan, of the University of Chicago, and one of the leading figures in the study of climatic change.

What *is* controversial is precisely how much of the Earth's surface temperature will rise with a given increase in greenhouse gases. Complications arise due to the process known as **feedbacks**. For example, if adding more greenhouse gases to the atmosphere increases the surface temperature, some of the Earth's ice cover will melt. The bright surface of the ice, which reflects heat back into space, will be replaced by the darker blue of the oceans or the green of vegetation, which absorb more heat. An increase in the amount of cloud brought about by higher air temperatures evaporating more water will provide another feedback mechanism, because clouds both *trap* heat rising from below and *reflect back* sunlight falling on their upper surfaces. To understand how the greenhouse effect works, and how it relates to the thinning of the ozone layer, we must take a closer look at the atmosphere.

The life shield

The Earth's atmosphere can be likened to the skin on our bodies. It protects us, regulates our (global) body heat, and permits the controlled passage of substances between the body and its environment. It is a sort of life shield which, like our skin, is vulnerable to damage. The increasing concentration of greenhouse gases is like a thickening of the skin, a change which may make us uncomfortably warm all over. The thinning of the ozone layer is like a raw patch scraped off our skin, removing some of our protection. With typical human perversity, we are damaging ourselves in both ways at the same time.

In reality, the atmosphere is a thin mantle of gases, mainly nitrogen (which makes up more than 78 per cent of the volume), and oxygen (which makes up nearly 21 per cent). It reaches to a height of about 1,000km above the Earth's surface. Scientists divide it into four zones, according to certain characteristics. In ascending order, these zones are the **troposphere**, **stratosphere**, **mesosphere** and **thermosphere** (see fig. 3). About three-quarters of all atmospheric gas is compressed into the troposphere, which extends to a height of about 8km above the poles and 17km over the equator. The air temperature in the troposphere drops steadily from an average of about 15°C at sea level to about minus 53°C at the boundary with the stratosphere, which is called the tropopause. The barometric pressure, of around 1,000 millibars in the lower troposphere on our TV weather maps, falls to a quarter of this level at the tropopause. The difference in pressure and temperature are important influences on the weather.

In the stratosphere, which ends at some 50km above the Earth's surface, the temperatures rise again with height to about 10°C. In the thin air of the two higher layers the temperature inversion is repeated, falling to about −90°C in the mesosphere and rising to 1,200°C in the thermosphere.

The atmosphere performs three roles which are vital to life on Earth and to the story of climate change:

☐ it keeps the Earth warm;
☐ it takes part in a constant exchange of chemical elements and compounds with the seas, the soils and the living matter of the Earth;
☐ it generates our weather.

All three functions are linked to one another and to chemical and biological processes taking place on the land and in the seas. By changing the atmosphere we are throwing all these interrelated activities out of kilter.

Mixed with the oxygen and nitrogen in the atmosphere are water vapour, which has a highly variable concentration of up to 3 per cent; fine particles of dust and chemical compounds, known as aerosols; and thin traces of many other gases. Some of these trace gases occur in such minute quantities that it's only recently that scientists

The Atmosphere

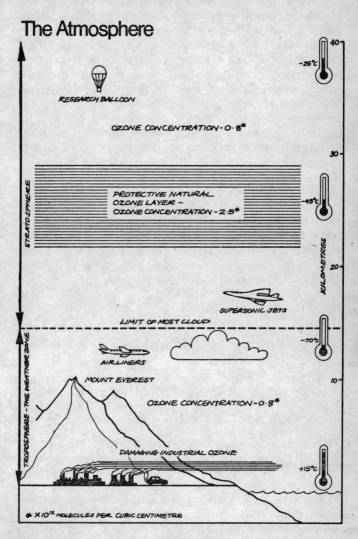

RESEARCH BALLOON

OZONE CONCENTRATION - 0·8*

PROTECTIVE NATURAL
OZONE LAYER -
OZONE CONCENTRATION - 2·5*

SUPERSONIC JETS

LIMIT OF MOST CLOUD

AIRLINERS

MOUNT EVEREST

OZONE CONCENTRATION - 0·8*

DAMAGING INDUSTRIAL OZONE

STRATOSPHERE

TROPOSPHERE - THE WEATHER ZONE

KILOMETRES

40
-25°C

30

-45°C

20

-70°C

10

+15°C

* × 10¹² MOLECULES PER CUBIC CENTIMETRE

Fig. 3

have been able to develop instruments sensitive enough to detect and measure them. Some trace gases play a role which is out of all proportion to their volume.

One such trace gas is **ozone**, a variety of oxygen with three atoms instead of two in each molecule. The special property of ozone is that it absorbs most of the sun's **ultraviolet radiation** – the invisible pulses of energy which can damage animal and plant life by altering the structure and operation of DNA molecules (the genetic material which control the formation of new cells). Ozone is found throughout the atmosphere but most of it is concentrated in the stratosphere between 20km and 50km above the Earth's surface.

Several other gases have what is, in effect, the reverse property. They let the sun's visible rays pass through them to strike and warm the Earth's surface, trap the heat which returns to the atmosphere as invisible infra-red radiation. The glass in a greenhouse works the same way: it is transparent to incoming visible sunlight but absorbs the infra-red rays, thus retaining their heat.

A planet fit for life

The human species, aided by the industries at its command, has significantly altered some of the planet's major chemical cycles. We have increased the carbon cycle by 25 per cent, the nitrogen cycle by 50 per cent, and the sulphur cycle by over 100 per cent. We have increased the flow of toxins into air, water, and food chains. We have reduced the planet's green cover, while our factory outpourings reach the upper atmosphere, and far into the oceans. And as our numbers grow, so will these perturbations.

James Lovelock, *Gaia Atlas for Planet Management*

Life is full of curious contradictions and paradoxes. This book is about the way we are threatening our future

The heat budget

The heat of the sun is the source of all our energy, driving a multitude of processes from chemical reactions to the movement of the winds. Every bit of sunlight absorbed by the Earth and its plants, oceans and atmosphere, is used to keep the machinery of life in motion and is then discarded into space. This ebb and flow of heat maintains an even, comfortable global temperature which is disguised by the contrasts of winter and summer, of equatorial noon and polar night.

The Earth's mean surface temperature, which varies slightly over long periods, is now about 15°C. It has been remarkably constant over hundreds of thousands of years. The warmest period in the last 100,000 years was only 1°C hotter than now. The regular swing between ice ages and warmer, intervening periods – called inter-glacials – such as the present, has been no more than 5°C. The global temperature when life began was perhaps no more than 7°C warmer than now. The greenhouse effect is the main element in this balancing act. Without it the planet would be frozen solid.

The amount of solar radiation arriving at the outer surface of the atmosphere varies according to the season and the time of day, and on a longer cycle is determined by the Earth's orbit of the sun. These variations, particularly the longer-term ones, are of some importance to scientists studying climate change because they can mask the effect of changes in the atmosphere. There are also considerable variations in the amount of solar energy absorbed by various surfaces. Highly reflective surfaces like the tops of clouds, snow and ice, bounce much of the sunlight which falls on them straight back into space. The proportion depends on the reflective quality of the surface, which is called the **albedo**, or whiteness. Clouds reflect between 20 per cent and 70 per cent, freshly fallen snow as much as 95 per cent, dense forests as little as 5 per cent. In this way, about 30 per cent of solar radiation is rejected by the Earth.

The capacity of different types of the Earth's surface to absorb heat also varies. Water, in particular, needs much more energy to raise its temperature than a comparable amount of sand or vegetation, and is much slower in releasing its heat back into the atmosphere when the surrounding temperature falls. Over long periods of time the heat accumulated in the surface water of the oceans is slowly carried down by currents to the depths. These factors make the oceans highly efficient global storage heaters. The heat given off by the oceans and land surfaces as infra-red radiation would drift away into space were it not trapped, and then re-released in all directions, by the greenhouse gases.

The Earth absorbs from the sun, and discards into space, equal amounts of energy, maintaining a relatively stable global surface temperature. There are, however, variations in the balance of gain and loss at different parts of the globe. The equatorial zone is a net importer of solar energy and polar regions are net exporters. If no other factors were in play, the tropics would be much hotter and the poles much colder than they actually are at present. But heat always moves from a hotter to a colder body. This fundamental law of physics sets the whole of the atmosphere in motion, mixing its gases, distributing its heat, creating the winds and the weather. Weather patterns throughout the world are connected to one another by this general circulation of the atmosphere.

Heat is also moved from the tropics to the poles by the general circulation of the oceans. A further distribution of heat takes place through the evaporation and precipitation of water. The complex interaction of all these processes make it beyond the present ability of scientists to accurately model the climatic consequences of the greenhouse effect.

Predictions of global warming made in the 1960s caused much confusion because there were, at the same time, predictions that the Earth was moving towards another glacial period. Since an actual reduction in global average temperatures was then occurring, the latter prediction seemed more credible. In fact, climatologists maintain there is nothing incompatible about a medium-term global warming due to greenhouse gases, and a longer-term cooling caused in the main by a period change in the Earth's position relative to the sun. Estimates place the start of the next ice age at 1,500 to 5,000 years in the future.

health, comfort, and perhaps our very survival, by thinning the protective layer of ozone and thickening the mantle of greenhouse gases. Yet in so doing, we are very slowly returning the atmosphere to an earlier state in which greenhouse gases predominated and there was no ozone. It was because of that topsy-turvy situation, some 3,500 to 4,000 million years ago, that life on Earth began.

The first signs of life on Earth apppeared when the atmosphere consisted largely of water vapour and carbon dioxide, both greenhouse gases, which were brought up from the planet's hot interior by volcanoes. Today, such a thick blanket of greenhouse gases would make the Earth intolerably hot. Then, however, the sun was some 30 per cent cooler than it is now. With a greenhouse blanket not thicker than today's, the planet would have been locked in ice. The thicker blanket of that era trapped just enough warmth to let the water vapour condense and form oceans.

The early oceans were rich in nitrogen, sulphur and carbon, the basic chemical components of life. These reacted with one another to form complex molecular compounds. No one can be quite sure how, among all this random chemical activity, the right combinations

occurred to create the first seeds of life, cells capable of
reproducing themselves. Two forms of intense energy
capable of altering molecular structures were available to
provide the necessary jolt: ionising radiation from radioactive materials in the Earth's crust, and ultraviolet radiation from the sun.

The process of evolution, which transformed some of
these primitive cells into more complex and diverse life-
forms, also modified the composition of the atmosphere.
Primitive cellular life fed entirely on the 'soup' of organic
chemicals from which it sprang and in which it lived. As
this source of free nourishment was used up, free-
floating algae-like plants learned to use sunlight to
manufacture their nutrients from the carbon and water in
the atmosphere. This is the process called **photosynthesis** which releases oxygen as a waste product. The
build-up of atmospheric oxygen had two effects. Molecules of oxygen (O_2) are broken apart by sunlight and
some of the atoms reassemble as ozone (O_3). Ozone
blocks the sun's ultraviolet radiation, allowing more
complex life-forms, which cannot tolerate that level of
solar interference, to develop and survive. Some of these
life-forms acquired another means of making use of the
atmosphere, breathing in oxygen and exhaling carbon
dioxide. Thus the first animal life appeared.

While terrestial life – the biosphere – has grown more
diverse and abundant, and the composition of the atmosphere has changed, the surface temperature of the
planet has remained remarkably constant. There have
indeed been warmer and cooler periods but the changes
have been small and slow. Larger, more violent changes
which might easily have happened as the sun grew
warmer, the atmosphere altered and life became more
profuse, would almost certainly have brought life to an
end. By the same token, life as we know it now could not
survive if all the oxygen or carbon in the atmosphere was
used up. Some mechanism seems to be at work keeping
everything in balance.

Jim Lovelock, an independent-minded but much respected English scientist, has devised an explanation of this mechanism which he calls the Gaia Hypothesis, after the Greek name for Mother Earth. Lovelock (whose description of life's origins we have drawn on) argues that Gaia behaves like a single complex living organism. It is capable of making the necessary adjustments to the composition of its components – the water, air and living matter – and to the exchange of materials between them, to maintain the right conditions for the planet as a living system to evolve and survive. 'The atmosphere is not merely a biological product,' Lovelock says, 'but more probably a biological construction – like a cat's fur, or a bird's feathers; an extension of a living system designed to maintain a chosen environment.'

Whether the adjustments are in fact made by some kind of automatic self-correcting mechanism, or by blind chance, the result has been a kindly planet, fit for life. But what we are doing to the atmosphere now is threatening to upset that precarious balance.

Scientists, philosophers and ordinary men and women have been studying and analysing the basis and meaning of life as far back as records exist. Remarkably for such a curious species, however, it is only in the past three decades that humans have made a serious global effort to measure the inter-related systems and atmospheric gases of their planet.

Enter the scientists

Every great advance in science has issued from a new audacity of imagination.

John Dewey, *The Quest For Certainty*

In 1957 the eyes of the world were on the British Commonwealth Trans-Antarctic Expedition: two converging teams led by Vivien Fuchs and Edmund Hillary making

The carbon cycle

Carbon is a basic building block of life. It is found in
rocks, soil, water, plant and animal life, and the atmosphere. Plants use photosynthesis to get the carbon
they need to build their cells from the atmosphere.
Animals get theirs by eating plants and other animals.
Carbon is returned to the atmosphere as carbon dioxide when plants are burned or decay, and when
animals breathe.

Atmospheric carbon dioxide is also absorbed by the
oceans. Some of it is returned to the atmosphere. The
rest is carried slowly to the ocean depths, both directly
by the physical movement of the waters and indirectly
in the remains of minute sea creatures. A small proportion of this solidified carbon is locked into the
sediment at the bottom of the ocean.

On land, the sedimentary rocks of earlier seas
slowly yield their carbon (and other elements) to
underground water, streams and weathering, and
return it to the seas. Other ancient deposits of carbon,
the remains of plants and animals in the seas and on
land, are the fossil fuels – coal, oil and gas – which we
burn today. By doing this, we are returning to the
atmosphere energy which has been safely locked
away for millions of years.

The oceans, the atmosphere and the biosphere are
what scientists call **sources** and **sinks** of carbon. As
long as the transfers from sources to sinks remain in
balance, the atmosphere and biological processes continue undisturbed.

Harvard climatologist Michael McElroy explains:

A carbon atom released to the atmosphere will flit
back and forth between plants, soil, air and water
for approximately 100,000 years before eventually
returning to the relatively quiescent reservoir of the
sediments. The average carbon atom has made the
cycle from sediments, through the more mobile

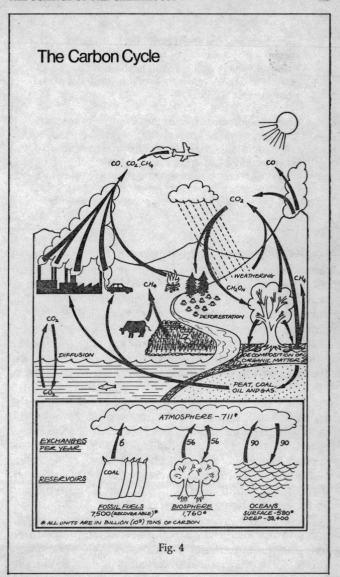

The Carbon Cycle

Fig. 4

compartments of the Earth, back to sediments, some twenty times over the course of the Earth's history. Similar cycles regulate the distribution of other life-essential elements such as nitrogen, phosphorous and sulphur.

Carbon dioxide molecules in the atmosphere are birds of passage, en route between the seas and vegetation. But they do not, so to speak, just fly around waiting to be absorbed by a leaf or a wave. They pay their way by acting as a greenhouse gas, maintaining a temperature without which the exchanges of carbon dioxide between water, air and vegetation could not take place.

the first land crossing of Antarctica. It was the most newsworthy event in a world-wide programme of scientific research known as the International Geophysical Year (IGY). The IGY, which involved 30,000 scientists from sixty-six countries at more than 1,000 research stations, was described by one newspaper as 'the most significant peaceful activity of mankind since the Renaissance'. Thirty years on we can recognise IGY as the turning point in climate research. It marked the start of the British Antarctic Survey's ozone monitoring, which led in 1985 to Joe Farman's discovery of the hole in the ozone layer. It began the programme of studying the Earth and the atmosphere from space, using rockets and balloons, which has added enormously to scientific understanding of climate and climatic change. As a late addition to the programme, it saw the establishment of atmospheric monitoring stations at the South Pole and on the Mauna Loa volcano in Hawaii.

These sites were chosen because they were far removed from localised air pollution, which would interfere with measurements of trace gases in the atmosphere at large. The Mauna Loa observatory is described by its director Elmer Robinson as 'one of the most favourable and accessible locations for measuring undisturbed tropospheric air'.

The sites were established at the instigation of the Scripps Institution of Oceanography in California which produced, in 1957, the first report to put global warming firmly on the scientific agenda. Until then, most climatologists had dismissed the warnings of Arrhenius and his followers, arguing that the extra carbon dioxide produced by industry would be absorbed by the oceans. The Scripps team, headed by Roger Revelle, concluded that half of it would stay in the atmosphere. 'Mankind is now engaged in a great geophysical experiment,' his report warned, in words that were to be adopted and paraphrased by successive waves of scientists, administrators and politicians as they too became convinced that the greenhouse effect would overheat the planet.

The first readings in 1958 from Mauna Loa showed that carbon dioxide was mixed with other gases in the atmosphere at a concentration of about 315 parts per million by volume (ppmv). This compared with a concentration of about 270 to 280ppmv in 1850. The 1850 figure, and those for other periods before actual atmospheric measurement began, have been reconstructed from bubbles of ancient air trapped in the ice sheets of Antarctica and Greenland, and other datable atoms of carbon locked in the annual growth rings of trees and in lake sediments. By analysing the radioactive carbon-14 isotope in any carbon sample, scientists can tell the sample's age and whether it came from burning fossil fuel or natural sources. They can also provide a reliable record of how carbon levels have changed with time. The same technique is used for dating archaeological finds.

Successive measurements at Mauna Loa show the concentration of the atmosphere steadily rising to a present figure of about 350ppmv (see fig. 3). There is a regular annual rise and fall, amounting to about 12ppmv, which reflects an increased intake of carbon dioxide by vegetation in the spring and summer, and an increased output from decaying plants and the soil in the autumn and winter. But the relentless upward swing is now rising at the rate of about 1.5ppmv a year. Such evidence points convincingly to a man-made addition to the total volume of carbon dioxide which is unbalancing the natural exchange between the atmosphere, the biosphere and the oceans. The concentration has increased by about 25 per cent since the industrial revolution, half of the rise accountable for in the last thirty years. The increase, although large, is not enough to account for all the carbon dioxide added to the atmosphere by human activity. Some of it, 40 to 50 per cent, has been removed. Climatologists assume that this lost fraction has been absorbed mainly by the oceans, and perhaps partly by vegetation. The precise reason is unknown.

The rapid increase in concrete evidence as to what was happening in the atmosphere, which started in the IGY, was complemented by the arrival of increasingly powerful computers. These enabled scientists to run complex mathematical models describing the behaviour of the atmosphere and its likely response to human interference. This in turn led to a rapid explosion in research and some increasingly worrying results. The models showed that doubling the concentration of carbon dioxide from its relatively stable pre-industrial level eventually was likely to bring a several degree increase in the mean surface temperature of the Earth. The calculations took account of some of the feedback mechanisms described earlier – the reduced reflection of the sun's energy (called the albedo) due to less ice and snow, and the increase in water vapour which doubles the greenhouse effect of carbon dioxide. Because energy analysts in the early 1970s were forecasting that the world would go on consuming fossil fuels – the main source of extra carbon dioxide – at a rapidly increasing rate, the atmospheric models showed that the doubling would happen quite quickly. It was not a matter of centuries, as Arrhenius had predicted, but of decades.

Dirtying the window

We are conducting a giant experiment on a global scale by increasing the concentration of trace gases in the atmosphere without knowing the environmental consequences.
World Meteorological Organisation, 1985

Global warming is usually calculated in terms of a doubling of the atmospheric concentration of carbon dioxide. However, in the last few years scientists have discovered that other gases come into the picture, and in a particularly disturbing way. Together, these trace gases now account for up to half the predicted global warming effect of greenhouse gases (see figs. 5 and 6).

One of these gases is **methane** (CH_4) which contains carbon, and, like carbon dioxide, is emitted to the at-

Relative Contributions to the Greenhouse Effect in the 1980's

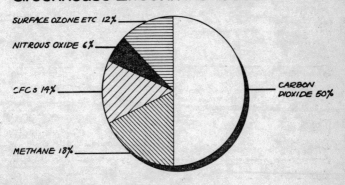

SURFACE OZONE ETC 12%
NITROUS OXIDE 6%
CFCs 14%
CARBON DIOXIDE 50%
METHANE 18%

SOURCE   J. HANSEN ET AL, 1988

Fig. 5

The Greenhouse Gases

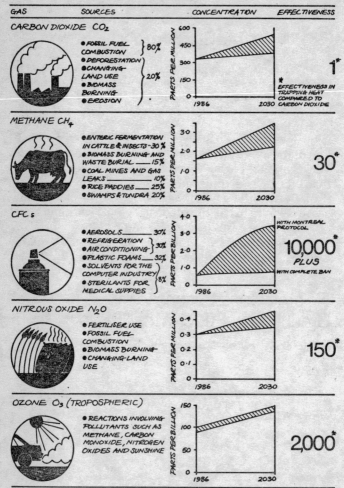

GAS	SOURCES	CONCENTRATION	EFFECTIVENESS
CARBON DIOXIDE CO_2	• FOSSIL FUEL COMBUSTION } 80% • DEFORESTATION • CHANGING LAND USE } 20% • BIOMASS BURNING • EROSION	PARTS PER MILLION (0–600) 1986 → 2030	1* * EFFECTIVENESS IN TRAPPING HEAT COMPARED TO CARBON DIOXIDE
METHANE CH_4	• ENTERIC FERMENTATION IN CATTLE & INSECTS – 30% • BIOMASS BURNING AND WASTE BURIAL ___ 15% • COAL MINES AND GAS LEAKS ___ 10% • RICE PADDIES ___ 25% • SWAMPS & TUNDRA 20%	PARTS PER MILLION (0–3·0) 1986 → 2030	30*
CFCs	• AEROSOLS ___ 30% • REFRIGERATION • AIR CONDITIONING } 30% • PLASTIC FOAMS ___ 32% • SOLVENTS FOR THE COMPUTER INDUSTRY • STERILANTS FOR MEDICAL SUPPLIES } 8%	PARTS PER BILLION (0–4·0) WITH MONTREAL PROTOCOL WITH COMPLETE BAN 1986 → 2030	10,000* PLUS
NITROUS OXIDE N_2O	• FERTILISER USE • FOSSIL FUEL COMBUSTION • BIOMASS BURNING • CHANGING LAND USE	PARTS PER MILLION (0–0·4) 1986 → 2030	150*
OZONE O_3 (TROPOSPHERIC)	• REACTIONS INVOLVING POLLUTANTS SUCH AS METHANE, CARBON MONOXIDE, NITROGEN OXIDES AND SUNSHINE	PARTS PER BILLION (0–150) 1986 → 2030	2,000*

Fig. 6

mosphere by natural as well as man-made processes. Methane – also known as marsh gas – is produced by bacteria which live in an environment free of oxygen: at the bottom of marshes and rice paddies and in the guts of animals, particularly ruminant, or cud chewing, species. In the ruminants, which include cattle, sheep and camels, the bacteria are active in the fore-stomach, breaking down the otherwise indigestible cellulose in grass. Cattle emit methane at quite remarkable rates. A typical domestic cow produces about 200 litres of methane a day, compared with about 12 litres of milk. Cattle numbers more than doubled between 1960 and 1980, creating a huge army of mobile polluters. In some other animals, including horses and rabbits, and in some insects, noteably cockroaches and termites, methane is produced in the lower intestine. About one-third of humans are said to have methane-producing bacteria in their systems.

Methane is also produced by rotting organic matter in refuse tips, and when wood and other vegetation is burned. There are leakages from gas pipes, coal-mines and other underground sources. As global warming occurs, large quantities of methane trapped in the frozen tundras of the North may be released to add to the atmospheric blanket and reinforce the greenhouse effect. Recent research suggests that there is methane trapped in geological formations on the continental shelves of the oceans which could also be released by temperature rises.

The amount of methane released to the atmosphere is hard to assess. Atmospheric monitoring, which began in 1978, and ice-core evidence suggests that the concentration has increased from 0.65 parts per million to 1.7 parts per million over the past 200 years, with a current rate of increase of about 1 per cent a year. Unlike carbon dioxide, methane is a chemically reactive gas which is broken down in the atmosphere. The products of these reactions include other greenhouse gases – carbon dioxide and water.

Another greenhouse gas which comes from both natural and man-influenced sources is **nitrous oxide** (N_2O), once used as a dental anaesthetic and nicknamed 'laughing gas' because it is mildly intoxicating: hence the grim joke that if the greenhouse effect proves fatal we shall at least die happy. Nitrous oxide is part of the natural life cycle, being both produced and consumed by biological processes in soil and water. Man-made sources, estimated to represent about 45 per cent of the output to the atmosphere, are principally fossil fuel combustion, the cultivation of the soil and the use of nitrogenous fertilisers, biomass burning (see fig. 6), and animal and human wastes. Present atmospheric concentrations are much lower than those for carbon dioxide – about 305 parts per billion (ppbv), rising by about 0.2 per cent a year. Evidence from ice cores suggests a pre-industrial level of around 285ppbv.

Ozone is a naturally occurring gas which is found in the atmosphere. It is also manufactured industrially as a purifying agent. It is created in two distinct ways, and plays two separate roles. In the upper atmosphere, where the greatest concentration is to be found, it is generated by the action of solar radiation on oxygen molecules. The ozone in turn absorbs ultraviolet radiation, breaking down into free oxygen atoms which recombine as oxygen and ozone. The reaction warms the stratospheric air. Some ozone molecules descend from the stratosphere to the troposphere, but the concentration there is generally much lower. In the air close to the Earth's surface, ozone is created by the action of sunlight on fossil fuel pollutants – hydrocarbons and oxides of nitrogen – from cars and other sources. The resulting photochemical smog is damaging to plant and human health. Tropospheric ozone also acts as a greenhouse gas.

Carbon dioxide, methane and nitrous oxide are all naturally occurring gases whose abundance in the atmosphere is being increased by human activity. The remaining greenhouse gases are human artefacts, designed for our convenience. These are the **chlorofluorocarbons** (CFCs), halons, and related compounds. CFCs were developed in 1930 as a safe cooling agent for refrigerators and air conditioners. (Their inventor, Thomas Midgely, of General Motors, also invented leaded petrol.)

The attraction of CFCs is that they are chemically stable and inert – in other words, they don't readily break down or react with other chemicals or metals. Initially, CFCs appeared to be both useful and harmless, a triumph of the industrial chemists' art. They were adopted as the propellant in spray cans, as the blowing agent for making plastic foams, and as a solvent to clean electronic components. Halons are used as the propellant in fire extinguishers. A related substance, methyl chloroform, is a dry cleaner and degreasing agent. It was introduced to replace a similar but more volatile substance, trichloroethylene, which contributes to the formation of photochemical smog. Used as propellants and foam-blowers, CFCs are released directly to the atmosphere. They also leak during manufacture, from industrial (not domestic) refrigerators and air conditioners, particularly during servicing, and from the open tanks used in cleaning electronic circuit boards.

The CFCs are now notorious for their destruction of stratospheric ozone. They are also extremely effective greenhouse gases. Although their atmospheric concentration is minute, measureable only in parts per trillion (one trillion = one million million), they block much more heat than carbon dioxide and water vapour. The atmospheric concentrations were first measured by Jim Lovelock in the early 1970s. Having invented an extremely sensitive instrument, an electron capture detector, Lovelock hitched a ride on a sea voyage from Britain to Antarctica and back to try it out. At that time he regarded their presence in the atmosphere as harmless and useful for tracing the movement of air masses. Now he argues that halting their use would be 'the one single step we could take immediately that would make a strong positive contribution to reducing the greenhouse effect.'

Compared with carbon dioxide and water vapour, all the other greenhouse gases are much less abundant. Collectively, however, they contribute as much to the warming effect because carbon dioxide and water vapour block only *some* of the wavelengths in which infra-red radiation travels. They leave a 'window' through which 70 to 90 per cent of the Earth's infra-red radiation escapes to space. Methane, nitrous oxide, ozone and the CFCs absorb radiation in the otherwise unblocked wavelengths and thus 'dirty' the window. Compared with a molecule of carbon dioxide, a molecule of either of the two most abundant chlorofluorocarbons, CFC11 and CFC12, is more than 10,000 times as effective in trapping heat. Ozone is about 2,000 times as effective, nitrous oxide about 150 times, and methane about 30 times. They also have a greater life span. Methane molecules survive for 10 years, CFCs for 70 to 110 years, and nitrous oxide for 170 years. The extremely persistent nature of CFCs is one of the main reasons behind the push by scientists and environmentalists to secure their complete ban.

Measuring the effect

When we put on more clothes we get warmer. If we put on too many we start to get uncomfortable. We begin to sweat, and perhaps to itch. Our hearts may begin to beat faster, and we may get short of breath. Our whole system becomes disturbed as it works overtime trying to restore its normal balance. Scientists call such an interference with the operation of a system a **perturbation** – a throwing into confusion or disorder. It is an evocative as well as a scientifically exact term.

Adding more greenhouse gases perturbs the atmosphere. It doesn't just make the air warmer, it creates

Stoking the furnace

The current global output of carbon dioxide from burning fossil fuels is just under 20,000 million tonnes a year (6 billion tonnes of carbon). The rate of output has grown on average 3 per cent a year since the middle of the nineteenth century, with a threefold increase taking place in the last forty years. These figures are derived from annual fuel production statistics. On average, every tonne of coal burned in a power-station or a domestic grate produces about 2.5 tonnes of carbon dioxide. Oil and gas produce less for the same amount of energy, by about 20 and 40 per cent respectively.

Some 85 per cent of carbon dioxide emissions come from the industrialised northern hemisphere. The USA accounts for 24 per cent. China produces 9 per cent but its contribution is growing rapidly, with a projected doubling of coal consumption in less than fifteen years. Britain produces 3 per cent, exceeded only by West Germany in the European Community. Australasia produces a mere 1.5 per cent while more than seventy lesser-developed countries (LDCs) together produce only 5 per cent.

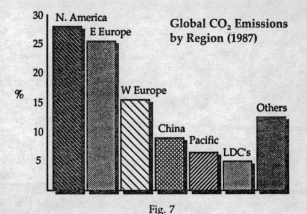

Fig. 7

While the industrialised world generates about 90 per cent of its energy from fossil fuels, developing countries rely heavily on wood and other forms of vegetation – known collectively as **biomass**. Burning biomass also produces carbon dioxide. Up to 1930, forest destruction was the main source of additional atmospheric carbon. Much of the destruction occurred in the Americas and northern Asia as forests were cleared for agriculture and settlements. Today, the same thing is happening in the tropical forests of South-East Asia, Africa and Latin America.

Destroying forests has a multiple and complex effect on the atmosphere. Carbon dioxide is released when the trees are burned, and more is released from the soil when it is tilled or otherwise disturbed. Because the vegetation has gone, less carbon dioxide is removed from the atmosphere. Loss of vegetation also increases the albedo (since forests are very dark and absorb a high level of the sun's energy), and reduces rainfall, with further potential effects on atmospheric behaviour.

Forests cover about 20 per cent of the world's land surface but produce 75 per cent by weight of all plant material – 950 billion tonnes a year compared with 7 billion tonnes from cultivated crops. Annual carbon dioxide emissions due to forest destruction and changing land use are estimated at 2.2 to 10.2 billion tonnes – 10 to 50 per cent of fossil fuel emissions. A figure of 4.6 billion tonnes is assumed by most climate modellers. If forest destruction in Latin America continues to accelerate at its present rate, the carbon contribution could increase to 40 to 70 per cent of estimated future fossil fuel emissions.

a more vigorous climate. Everything has to work harder to adjust to the extra pressure and restore the balance. Although climatologists talk about a 'giant experiment' with the planet, they cannot put the Earth in a laboratory, attach their instruments, add measured quantities of carbon dioxide to the atmosphere, and read off the results. They can only do this with models: not scaled-down physical replicas, but numeric models which mathematically reproduce the processes taking place in the atmosphere.

Modelling is a technique widely used in everything from designing ships to forecasting the behaviour of the economy. Although it is a complex exercise, it breaks down easily into two distinct parts. The first is the 'black box', the set of equations which attempts to reproduce all the forces at work in the real world system being studied. The second is the 'input', the information fed in by the modeller about something which is going to upset the way the system works. Neither of these components is completely reliable.

Models are less than perfect representations of reality and are thus an imperfect method of describing and predicting climatic change. This is partly because scientists do not yet fully understand how the atmosphere works and interacts with the oceans and biosphere, and partly because no computer currently available is powerful enough to carry out all the necessary calculations. Some elements, therefore, are missing altogether from the models, while others are represented by assumptions rather than facts, or included in an over-simplified form. The major deficiencies in climate modelling are the behaviour of the oceans and the role of clouds. Both are important as feedback mechanisms which modify the effect of greenhouse gases.

The oceans are a major source and sink of carbon, thus influencing the concentration of carbon dioxide in the atmosphere. One of the biggest conundrums in climate studies is that only half the extra carbon dioxide added by human activity remains in the atmosphere. It is assumed that the other half has been reabsorbed somewhere. Some of it may have been taken away by vegetation; most of it has probably been absorbed by the ocean, where it may be locked away for centuries. Scientists are unsure how much extra capacity the oceans have for absorbing carbon dioxide. Some suspect they may have just about reached their limit. Others think that they may be able to go on absorbing more carbon dioxide, keeping the amount left in the atmosphere in check. There is evidence that the carbon content of the oceans has varied, over long periods of time. When the world was cooler, during the ice ages, there was less carbon in the atmosphere, and more in the oceans, than there is now. Recent studies have also suggested that the present carbon content of the surface waters is much higher than had been generally assumed. But the evidence is skimpy. It is not clear whether the oceans develop the capacity to absorb more carbon because the climate changes, or whether the climate changes because the oceans start absorbing more carbon.

The oceans are also the Earth's main reservoir of heat. They absorb the sun's energy in large amounts and release it slowly, moving it around the globe. About half of the heat transferred from the equatorial regions to the poles is carried by the oceans. Climatologists have yet to develop adequate modelling techniques for either the physical or the biological roles of the oceans.

Clouds are important climate regulators because they both reflect the incoming rays of the sun back into space and trap the heat rising from the land and oceans. But the relative gains and losses of atmospheric heat involved in this process depend on the type and location of clouds as well as their size and frequency. Climatologists know that global warming will increase the amount of water vapour in the atmosphere, but they have not yet discovered a way of predicting cloud patterns in a more vigorous climate. Recent attempts at cloud modelling exercises have produced widely varying results, ranging from a doubling to a halving of global warming.

Inputs are the second area of error. Nobody knows the rate at which the world will continue adding carbon dioxide to the atmosphere. The best the modellers can do is draw up a series of scenarios based on various assumptions about the future rates of fossil fuel consumption. Typically, these range from a low scenario in which stringent efforts are made to reduce emissions, to a high scenario in which fossil fuel consumption rises in line with both a growing world population and expanding economies. Devising these scenarios involves a preliminary modelling exercise in which assumptions have to be made about population growth, economic growth, and the relationship between economic growth and energy consumption. The results of all these calculations are fed into models of the atmosphere to determine how long it will take for the carbon dioxide concentration to double, how much extra heat that will produce, and how long it will take the atmosphere, the biosphere and the oceans to adjust and produce a new equilibrium where energy flows are again in balance. Any temperature rise produced by doubling the level of greenhouse gases will not be immediately apparent. There is a time lag, due mainly to the slow reaction of the oceans, which is calculated at anything from fifteen to fifty years. Recognition of this delay is important, because if we wait for conclusive proof from air temperatures that global warming is occurring it will be too late to take action to reduce the subsequent twenty to thirty years of warming.

Calculating the effects of doubling the level of greenhouse gases is a somewhat artificial exercise. It simply provides a convenient cut-off point for modellers to produce comparable results. In reality, we can expect the level of gases to go on rising until we make a successful attempt to curb emissions, or to adjust the capacity of the biosphere to absorb excess carbon dioxide. Whether we can so finely tune our activities to achieve stable levels of greenhouse gases must be open to question. Unless mankind, or Gaia, can return the atmosphere to a state of stability we must be prepared for a perpetually bumpy ride.

Climate models

The most sophisticated climate change models, which mimic the atmosphere and the Earth's surface over the whole globe, are much the same as the models used by meteorologists to make weather forecasts – in other words, rather than predict what is going to happen in the atmosphere over the next few hours or days, the climate change models anticipate the next fifty or a hundred years.

However, there is one crucial difference. It is impossible, with the computer power currently available, both to calculate changes taking place over decades and to represent the atmosphere and the Earth's surface in the fine detail necessary for effective weather forecasting. So the greenhouse effect models cover the globe on a much broader scale, dividing the surface and the atmosphere into large boxes.

Typically, each box covers a surface area of 500 to 1,000 km sq and divides the atmosphere into ten or twenty layers. Each box gives an averaged-out description of the surface features and atmospheric conditions it contains. This means they cannot predict precise climatic changes for particular regions.

One 1,000 km sq box would cover almost the whole of the British Isles, averaging out conditions in Kent, Caithness and the Pennines, the Wash, Snowdonia and Galway Bay.

Principal modelling centres include the National Center for Atmospheric Research in Boulder, Colorado, the Goddard Institute for Space Studies (NASA) in New York, the Geophysical Fluid Dynamics Laboratory at Princeton University, the Soviet Hydro-Meteorology Centre in Moscow, and the Meteorological Office (the Met Office) in Bracknell, Berkshire.

Ironing out the differences between models and modelling techniques, so that results are more consistent with one another, is a key task in the process of reaching greater certainty about global warming. The job is being done by an international panel of climatologists chaired by John Houghton, director of the Met Office. The panel's report is due by the summer of 1990.

At the first meeting of the panel in January 1989 Houghton said:

> The end product will be the best statement we can give of how the climate is likely to change over the next century as a result of human activity. The task of the team will be to try to address the uncertainty and to see how we can quantify it in a useful way for people who have to make policy.

Warming the political climate

Increasing concern among scientists about the greenhouse effect has taken climatology from an academic backwater into the scientific mainstream. It has become part of a composite new science which brings together many traditionally independent disciplines. The US National Aeronautics and Space Agency (NASA) calls the new discipline **Earth System Science** and defines its goal as 'to obtain a scientific understanding of the entire Earth system on a global scale by describing how its component parts and their interactions have evolved, how they function, and how they may be expected to continue to evolve.' Sally Ride, one of NASA's former astronauts, calls its programme 'Mission to Planet Earth'.

Through the 1970s and early 1980s governments were sufficiently impressed by the arguments of climatologists to continue funding their research, but not so struck that they felt it necessary to sound alarm-bells or take practical action. While other atmospheric pollution problems, such as smog, acid rain, and lead – and latterly ozone depletion – attracted public concern and government action to reduce emissions, the greenhouse effect remained largely an issue for the scientific community.

This was certainly the case when the World Meteorological Organisation, an agency of the United Nations, convened the first World Climate Conference in Geneva in 1979. Kenneth Hare, who helped organise it, recalls 'sitting in an office in Geneva biting my fingernails because we could get little response from businessmen, engineers, doctors, farmers and fishermen; and little or none from politicians.' The conference drew up a World Climate Programme (WCP) in conjunction with the United Nations Environment Programme (UNEP), the International Council of Scientific Unions (ICSU) and other bodies. The WCP covers a wide range of activities, including: national and international research programmes on climate change and its impacts on the planet and society; the establishment of data banks and arrangements for exchanging information and assistance to countries to improve their climate research and monitoring capabilities. One of WCP's principle objectives is to 'warn governments of the potential economic, social and political impacts of significant climatic variations and changes, both natural and man-made.'

In October 1985, scientists from twenty-nine industrialised and developing countries met at a WCP workshop at Villach in Austria to review current research of the greenhouse effect. They reached sufficient agreement to draw some important conclusions:

☐ that the combined effect of the rising concentrations of all greenhouse gases might be the equivalent of a carbon dioxide doubling by the 2030s;

☐ that such a doubling might induce a global warming, perhaps some decades later, of between 1.5 and 4.5°C;

☐ that high latitudes would warm the most, especially in autumn and winter;

☐ that soils might become drier in the Northern Hemi-

sphere's mid-latitudes which include the world's chief wheat and corn belts;
☐ that sea levels might rise between 20 and 140cm.

The workshop argued that warming on this scale would have profound effects on global ecosystems, agriculture, water resources and sea ice. The scientists concluded with a blunt warning. Many important economic and social decisions on long-term projects – from irrigation schemes through drought relief arrangements, from coastal engineering projects to energy planning – were being based on the understanding that past climate was a reliable guide to the future. 'This is no longer a good assumption,' warned the WCP delegates.

Two further meetings were held by the WCP in 1987 – the first at Villach once again, the second at Bellagio in Italy. The Villach–Bellagio workshops were significant for two main reasons. Not only were global warming trends confirmed, but, for the first time, policies to respond to climatic change were discussed in detail and agreed. The final document, which set the agenda for the 1988 Toronto Conference, is reproduced in full in appendix 2.

The first Villach workshop was timely, for it took place at a point when experience of palpably extreme and damaging climatic events (such as the continuing African droughts and the repeated flooding in Bangladesh), began to converge with the results of scientific research.

In May 1985, scientists of the British Antarctic Survey, who had been watching a progressive reduction in the ozone layer over Antarctica during the Southern Polar spring, reported that concentrations had fallen by about 40 per cent. Such a shift, too large to be blamed on instrument error, was dramatic confirmation that something was destroying the stratospheric ozone layer.

In July 1986, climatologists from the Climatic Research Unit at the University of East Anglia (UEA) produced evidence that global temperatures had been rising over a

Into the ozone hole

The discovery in 1985 by Joe Farman and his colleagues in the British Antarctic Survey of the hole in the ozone layer, and subsequent evidence of a general thinning of ozone across the whole globe, played an important part in alerting both public and politicians to the related issue of climatic change. Farman, who has been taking ozone readings at the Survey's Halley base since 1957, is a passionate advocate of speedy action to halt CFC emissions. In March 1988, he told the UK House of Commons Environment Committee that current emissions would continue to damage the atmosphere for more than a century. 'The behaviour of ozone in Antarctica will not go back to the state it was in 1957 in my lifetime, or in your son's lifetime.'

Scientists began worrying about a possible loss of ozone in the early 1970s when it was predicted that supersonic passenger aircraft like Concorde, flying in the stratosphere, would produce enough water vapour and nitrous oxide from their exhausts to act as a catalyst in the destruction of ozone. Subsequent research suggested this fear was unfounded. Meanwhile, CFCs were identified as the real threat. CFCs are broken down by ultraviolet (UV) radiation in the upper atmosphere. Chlorine atoms liberated from the CFC molecules speed up the destructive action of UV radiation on ozone. A single chlorine atom can help destroy 100,000 ozone molecules.

The ozone hole, covering an area larger than the Antarctic continent, occurs inside the **polar vortex**, a strong pattern of circumpolar winds which effectively isolates the air inside from the rest of the atmosphere. The cold air, and icy clouds which form in the stratosphere, act like a kind of **chemical processor** producing the right mixture of chlorine atoms and nitrous oxide – another key factor – to ensure a rapid destruction of ozone when the sun returns in the spring.

Farman's discovery prompted a fierce academic

debate about whether CFCs or a violent change in the air circulation patterns around the Pole was responsible for the damage. An urgent reappraisal of existing data revealed that the hole would have been discovered earlier if the computer analysing US satellite observations had not been programmed to exclude very low readings as misleading anomalies. It also showed a downward trend in ozone concentrations globally, amounting to a 3 per cent decrease since 1969, with unexpectedly large depletions occurring in middle and high northern latitudes in winter. The changes are greater between latitudes 53 and 64 degrees North, which includes a large part of the UK.

A major scientific campaign took place in the Antarctic spring of 1987, using two aircraft based at Punta Arenas in Chile. The Antarctic Airborne Ozone Experiment coincided with the deepest and longest-lasting ozone hole yet recorded. Ground and airborne observations revealed that the total column of ozone within the hole was depleted to about 40 per cent of its pre-1975 level, and that there was a pocket of air at an altitude of 14 to 18km in which it was reduced by 97 per cent.

Farman told the Commons Committee that the expedition ended the debate as to whether the chlorine from CFCs was affecting ozone. But it left open the question of how far the ozone destruction might go. Models had failed to predict the hole. Farman concluded that such information was crucial:

> For heaven's sake, if we fail to do that, it shows how little we can assess these risks. I am afraid I can offer you no hope whatsoever on the scientific evidence that there will not be a global depletion of ozone of the order of 15 to 20 per cent in the next ten years.

It later emerged that as the 1987 hole broke up, it drifted across southern Australia and New Zealand. Over Melbourne, the ozone level fell for a few days in

December by almost 12 per cent, allowing higher levels of UV-B radiation to reach the Earth's surface. Australia already has the highest incidence of skin cancers in the world which are caused by exposure to UV radiation.

The 1988 ozone hole was less intense. Farman told a London conference in December 1988 that there appeared to be a regular cycle of more and then less intense reductions in the ozone over the Antarctic. He believes that in years when the hole is deep and long-lasting, relatively little chlorine-rich air is able to reach the Pole, so that there is less ozone destruction in the following year. Less ozone destruction leads to a less intense, shorter lasting hole which allows more chlorine-laden air in to repeat the cycle.

Although the Arctic region also develops a polar vortex, it is often disrupted by an invasion of warm air, so the same conditions are not likely to develop. However, there is evidence of some accelerated ozone depletion in the Arctic, which was tested in 1989 by an international investigation on the same lines as the Antarctic Airborne Experiment of 1987. As this book went to press, initial results from the experiment showed chlorine levels fifty times greater than predicted. NASA's Robert Watson told reporters in Washington that the results would 'present world governments with firm evidence that firm global action must be taken.'

period of 134 years, and that the three warmest years had occurred in 1980, 1981 and 1983. Three months later a team from the US Geological Survey reported that frozen earth beneath the Arctic tundra in Alaska had warmed between 2.2 and 3.9°C in the past century – further evidence that the world was hotting up.

The UEA team, with other scientists from both the UK Meteorological Office and from America, updated the temperature series in April 1988 and revealed that 1987 was the warmest year recorded. Three months later, while the US sweated and worried its way through its worst drought for fifty years, Jim Hansen of the Goddard Institute for Space Studies (GISS) in New York took his latest modelling results, which included a predicted increase in summer heat waves, to a congressional hearing. He told the politicians that the global temperature up to 1 June 1988 was substantially warmer than the similar period in any previous year on record:

> We can state with about 99 per cent confidence that current temperatures represent a real warming trend rather than a chance fluctuation . . . It is time to stop waffling so much and say that the evidence is pretty strong that the greenhouse effect is here.

The UEA researchers subsequently confirmed that 1988 had been even warmer than 1987, making the 1980s the warmest decade since records began. The world suddenly started to look up and take notice.

A model warning

Ozone depletion and global warming are distinct but not entirely separate events. They are linked by the dual roles of CFCs as ozone-eaters and global warmers, and by the fact that ozone depletion lets more solar radiation through to warm the Earth and reduces temperatures in the stratosphere. A measured fall in stratospheric temperature is regarded as one of the key indicators that global warming due to the greenhouse effect – rather than some other cause such as a variation in the sun's heat – has begun. The evidence for such a change is not yet complete. Measurements since 1979 show a temperature reduction of about 1.5 to 2 °C in the stratosphere at 40km above the equator. According to Bob Watson, who chairs the international panel of scientists reviewing evidence on ozone depletion, this is consistent with the measured losses of ozone and increases in greenhouse gases.

The ozone story holds a significant lesson for all climate watchers or politicians determined to have absolute proof of global warming before they start to take action. Although ozone depletion from rising levels of CFCs was predicted, no model suggested, and no scientist imagined, that the Antarctic ozone hole would occur. Nor did models predict the lesser but still significant reduction of ozone at other latitudes. It took just nine years – from 1978 when the ozone hole was indiscernible, to 1987 when there was a localised loss of 97 per cent. Joe Farman of the British Antarctic Survey told the UK House of Commons Environment Committee.

> We have wiped out the heart of the ozone layer. This implies that there is a sort of threshold . . . and once you have got above that threshold, then things go. Our problem is that we do not know whether there is a critical threshold for greenhouse gases, above which 'things will go'.

3

LIFE ON A HOT PLANET

If current trends continue, the rates and magnitudes of climatic change may substantially exceed those experienced over the last 50,000 years. Such high rates of change would be sufficiently disruptive that no country is likely to benefit *in toto*.

The Changing Atmosphere – Implications for Global Security,
Toronto, 1988.

Leon Malard became famous in the summer of 1988. He was pictured on the front cover of *Time* magazine, baseball-capped, brawny-armed, sifting a handful of dust and dried barley stalks against a lurid sky on his North Dakota farm. His story – and that of many other farmers, descendants of the pioneering sod-busters and the 'Oakies' who escaped the 1930s dust-bowl of the Mid-West to face the 'Grapes of Wrath' in California – filled many acres of American newsprint that summer. Most of the States, coast to coast, were gripped by the fiercest heat wave and drought for fifty years. There was growing public concern about global warming and it was a presidential election year.

Malard told *Time* that some of the land on his 1,200 acre farm was baked so hard he couldn't get his plough into it. 'The barley and oats are gone. The corn is beginning to turn white. The leaves are curling. If there is no rain, if the wind keeps blowing like this, if it stays so hot, all the

corn will be lost.' To the east, in Minnesota, where the
drought had been unbroken since 1986, things were
worse. Dean Hagan told reporters that when the thermometer topped 100°F for the third successive day, four
of his breeding sows 'just kind of gave up and died'. His
corn, instead of being as high as an elephant's eye, had
reached only thirty inches; his soy-beans stood at three
inches instead of three feet.

On the West Coast and in the Rockies, from the Ozark
Hills of Missouri to the woodlands of Georgia, there were
forest fires. 'From 25,000 feet above the southern flank of
Yellowstone,' wrote Michael White in *The Guardian* 'the
world's oldest national park looks much as Cologne or
Dresden must once have looked to Bomber Command.
Nearly forty fires are burning below, the smell of smoke
filling the aircraft as it moves downwind. Apparently the
smoke can be seen 1,600 miles away in Chicago.' At
Memphis, the Mississippi was at its lowest since records
began in 1872 and more than 3,000 river barges were
blocked by sand-bars. Severe water restrictions were in
force from Chicago to San Francisco.

America did not suffer alone in 1988. There was also
drought in the USSR; continued drought and unexpected
floods in Africa and India; floods and drought simultaneously in China; floods in Brazil, and Bangladesh; hurricanes in the Caribbean; a cyclone in New Zealand and a
typhoon in the Philippines. It surpassed 1987 as the
hottest year on record. Many hundreds died in the
storms and floods, many billions of pounds of loss and
damage were inflicted on man by the restless, unruly
atmosphere.

Climatologists were divided on whether man himself
should take the blame; whether, that is, these events
marked the onset of global warming. Jim Hansen of GISS
was almost certain that the greenhouse effect was upon
us. Others accused him of undermining the credibility of
science by crying wolf. The future will decide who was
right. For now, it is enough to look on these weather

Gilbert and Joan

Hurricane Gilbert hit Jamaica at 3p.m. on Monday 12 September 1988 and left a fifth of the island's 2.5 million people homeless. Crops of banana, coconut, sugar, coffee, vegetables and marijuana were destroyed, and tourist resorts were devastated. The then Prime Minister Edward Seaga declared that the burgeoning economy was set back a decade by the 'worst natural disaster in our modern history'.

Gilbert, a rare category 5 hurricane 'capable of doing catastrophic damage', killed at least a hundred people in the Caribbean and Mexico, and did billions of dollars worth of damage before blowing itself out in a series of tornadoes around the coast of Texas. Its winds reached 320km per hour.

The following month, Hurricane Joan brought a similar trail of death and devastation to Nicaragua. An estimated 70 per cent of trees were destroyed over an area of 15,000km sq. Gordon Hutchinson, a field worker for aid agencies, reported:

> What I saw resembled the well-known photographs of Hiroshima after the atomic bomb attack. From Santo Tomas to Bluefields virtually all the trees are down. From horizon to horizon all you can see is a light colour which is a combination of mud and vegetation burned by the over 200kph winds. The rivers are still in flood and are choked with mud and organic material. This is expected to kill off much of the river, coastal animal life and vegetation. People are still in a state of shock as could be observed by their vacant facial expressions. In the space of twelve hours they lost everything. There is serious concern at the possibility of disease. Water is either polluted or brackish.

Hurricanes, born in troughs of low pressure, nourished on heat, moisture and atmospheric disturbances, may become more frequent and up to 40 per cent stronger as global warming takes a grip.

extremes as a foretaste of what is to come, a series of practical lessons in living on a hotter planet.

The school of experience

When it comes to predicting how life on Earth will react to a warmer atmosphere, scientists have little to go on. There is some evidence, from ice cores, rocks, and fossils, of what the world was like during earlier warm periods, and of how it changed as the temperature rose and fell between and through the various ice ages. At the height of the last ice age, 18,000 years ago, large mammals like the mammoth and sabre-tooth tiger became extinct, and the now characteristic spruce and oak forests of Europe were absent.

By 10,000 years ago, oak and spruce were moving north, the type of ecosystems we know today had begun to appear, and the first primitive farming methods were being developed. Six thousand years ago, when summer temperatures were 2 to 4°C warmer than now throughout North America and Eurasia, spruce trees were spreading from Russia into Scandinavia, the prairies had reached their maximum extent in a drier North America, and strong summer monsoons raised lake levels in the now arid regions of Asia and Africa. The waters of Lake Chad were 110 metres higher than today. There were hippopotami and crocodiles in the Sahara.

Nearer at hand, there is evidence that a period of cooling which began around AD 1250 was responsible for the disappearance of the Norse settlements in Greenland. The European 'Little Ice Age' of 1500 to 1800 saw half the farms in Norway abandoned and the demise of cereal crops in Iceland.

The death of the mammoth, the northward migration of the pine, the southward retreat of the hippopotamus, and the contraction of the prairie were gradual changes on a planet where man's impact was still negligible. In none of those periods was the global mean temperature as high as it may become by the middle of the next century. To find a similar era, 3 to 4°C warmer than now, we have to go back some 3 to 4.5 million years.

Existing climatic regimes are another potential source of guidance. In theory, conditions in the South of France should give us some insight into what it will be like in southern England if, as predicted, temperatures there reach Mediterranean levels. But the mix of weather throughout the year which gives an area its characteristic climate depends on more than surface temperature. Its location, in relation to oceans and landmass and prevailing wind systems, is just as important. Local topography also helps to determine climate. Consequently, the value of simple area comparisons is limited.

We have to remember, also, that global warming will make the weather behave more violently and unpredictably. Past and present climatic norms may be imperfect guides. Recent experience of extreme weather events may prove more useful. Before we turn to what the experts predict will be the effects of climatic change fifty years or so from now, we should take a closer look at what happened in the USA in 1988.

Life in a dry season

The 1988 drought was caused by a ridge of high pressure lodged over the middle of the country, diverting the rain-bearing westerly winds to Canada and Mexico. It began early in the year, in April, and was compounded by several years of low rainfall in the south-east and the west. The snow on the Rockies, which usually fills the rivers in spring, lay less than half as deep as in a normal year.

Drought is a periodic hazard in the Great Plains of North America and the intensively farmed land is vulner-

The climatic cascade

Global warming is predicted by climatic modellers to increase the average temperature at the Earth's surface 1.5 to 4.5°C by the middle of next century, unless we substantially reduce greenhouse gas emissions. This rate of warming would be between 0.3 and 0.8°C a decade. Present concentrations of greenhouse gases have already committed the Earth to a further rise of 0.5°C or more on top of the 0.5°C already experienced. The average increase would be experienced in the mid-latitudes. In the Tropics, however, the effect would be less, perhaps as little as 0.1°C. Towards the Poles it would be greater, up to 8°C or more. There may be more warming in winter than summer. Increased evaporation would bring more rain and snow but some areas may be drier in summer. There would be more heat waves, droughts, storms and floods, and these are likely to be more severe. Seas would rise by 20 to 165cm, in response to expanding oceans and melting ice sheets.

The impacts of global warming would cascade through nature and society. The distribution of wildlife species would change and some species may become extinct. In the mid-latitudes, forests and areas suitable for growing staple crops would move towards the Poles, possibly by several hundred kilometres. Changes in the amount and distribution of heat and moisture will affect growth rates and yields, in some cases beneficially, in others adversely. Farming practices and incomes will change. Demand for water may increase but supplies may be reduced and have a greater risk of pollution. Energy needs will alter. Coasts will flood unless they are protected, and saltwater will infiltrate underground drinking water sources. Patterns of disease and mortality will change. Work and recreation habits may change. Investment and spending priorities for individuals, industries and public authorities will alter and the effects will ripple through national and international economies.

Climatic change, and the environmental and human responses to it, will interact through 'feedback' mechanisms. Changes in the amount and distribution of ice, snow, water and vegetation may modify the greenhouse effect for better or worse. The continued destruction of tropical forests, by human or natural means, and the melting of northern permafrost due to temperature rises could add significantly to greenhouse gases. Changes in the amount of energy used, and in industrial and agricultural practices, will alter the rate at which greenhouse gases accumulate in the atmosphere.

able to its effects. The drought of 1988 was comparable to the Dust-Bowl of 1934 and to other droughts of the early 1900s: worse in some respects, easier in others. Conditions have changed since the 1930s in ways which altered the impact and consequences. More effort is made today to prevent soil erosion, irrigation schemes are widespread and government aid is more readily available. Even so, the erosion was substantial, affecting some 5.25 million hectares. Three times as much land is irrigated today, but aquifers were drawn dangerously low in places. Irrigation equipment suppliers enjoyed a boom.

There is only a third of the number of farms now as in the 1930s, but they produce in a normal year five times as much corn and three times as much wheat. These days it is agricultural over-production, not scarcity, that is the problem in the US, as it is in most developed countries. The US crop surpluses help to feed the USSR and many developing countries, and supply much of the hard wheat used for British bread. The 1988 drought cut output by an estimated 31 per cent, leading to an increase in world prices, and causing many countries to draw heavily on their reserves of cereals. World grain stocks in store fell by 38 per cent to 286 million tonnes and world production dropped by over 8 per cent in two years. The US Agriculture Department said another drought in 1989 could produce catastrophic shortages.

Many beef and dairy cattle were slaughtered for lack of feed. It will take years for herds to be restored to their previous levels. In some places, cattle encountered another consequence of drought – feed crops containing dangerously high levels of nitrates which the water-starved plants had been unable to convert to protein.

Electricity output was reduced because water shortage affected hydro-electric plants and the cooling of nuclear-power plants. More coal, gas and oil were burned to make up for hydro-power losses. Because the Mississippi was running too low for the barges which normally carry large volumes of grain, coal and other bulk goods to the Gulf of Mexico, the railways flourished. 'This is probably the first time since the railroad was started in 1851 that we've made a profit in July,' said Harry Bruce, chairman of the Illinois Central Railways. Many farmers and farm suppliers, however, faced bankruptcy.

By late August, 66,000 fires had destroyed about 1.4 million hectares of forest. The traditional practice of letting fires burn to aid forest regeneration met with severe criticism. In Minnesota, an estimated 28 million young Christmas trees were killed or damaged. In many places trees suffered from water shortage and additional stress from ozone pollution and insect pests.

Water pollution problems worsened as the low river flows failed to dilute effluents. A $3.4 million scheme to counter the flow of saltwater up the Mississippi began. Fish populations were reduced and coastal oyster beds were endangered by increased salinity. There was more disease among wild animals. In parts of Montana, grasshoppers disappeared.

Thousands of extra human deaths were caused by heat stress. Hay fever, asthma and health problems related to air pollution were made worse by warm, stable atmospheric conditions. Farmers were at risk from silage gases and spore-laden dust. There were reported increases in spinal cord injuries caused by swimmers diving into shallow water and hitting the bottom.

Farming and food in 2050

A four-year study of the likely impact of global warming on agriculture in the Northern mid-latitudes by the International Institute for Applied Systems Analysis (IIASA), completed in 1988, confirms the relevance of the US drought to the hotter weather predicted by the climatologists. There is a crumb of comfort in what the IIASA forecasters are saying. At the lower end of the warming scale, the impacts they foresee are ones with which farmers and agricultural scientists should be able to cope, if changes in farming practices and food distribution are introduced. Farmers will have to modify what they grow and how they cultivate their land, but the impacts do not appear to go beyond the bounds of experience. At the middle to upper end of the possible warming scale, adaptation will be much more difficult. But many of the problems will be social, economic and political, rather than agricultural.

In some areas climatic change may help agriculture. Warmer weather will mean longer growing seasons and faster growth rates in the temperate and cooler latitudes, provided there is enough moisture available. The amount of rainfall, its distribution throughout the year, and the amount of moisture remaining in the soil – factors on which climate models are unreliable – will be critical. On the down side, soil fertility could change, and erosion increase. Where the new climate favours crops it also favours crop pests and diseases.

Higher levels of atmospheric carbon dioxide could, in theory, give a boost to output. Laboratory studies show that some crops and trees grow faster and bigger, and make better use of the available moisture, in an atmosphere richer in carbon dioxide. It remains to be seen whether this will happen in field conditions. Problems could occur where weeds respond more readily than the crops.

The climatic conditions suitable for various types of crop – wheat for example – are likely to shift towards the poles by some hundred of kilometres for every 1°C increase in temperature. The Canadian province of Saskatchewan, which produces 60 per cent of the country's wheat, will become more like the US state of Nebraska, with average temperatures 4.7°C higher by the mid-twentieth century, rainfall up by 15 per cent, and the growing season extended by seven weeks. However, droughts are likely to become more frequent and severe. Wheat production could fall by between 6 and 26 per cent. Farmers could reduce the losses by switching from spring to winter wheat – a change which has already been implemented in some areas. More than 7 million acres of uncultivated land in the Canadian north-west will have a climate suitable for arable farming, but the soils are so thin and poor that it may not be economic to cultivate them.

According to studies by the Environmental Protection Agency (EPA), on the US, the world's leading agricultural producer, there will be severe adverse effects in the east and south, balanced by improved growing conditions in the north and west. The country would be able to go on meeting its own food needs, even under the most extreme conditions predicted, but exports – critically important in meeting the food needs of other countries – may suffer. The Agency also predicts environmental costs:

☐ more competition for water resources;
☐ greater potential for ground and surface water pollution;

☐ increased use of pesticides to counter new pests;
☐ increased soil erosion;
☐ some loss of wildlife habitats.

Climate change, says the EPA, 'could exacerbate many of the current trends in environmental pollution and resource use from agriculture, and could initiate new ones.'

If the grain belts move north, the southern prairies might be better employed raising livestock. It has been suggested that they should be given back to the buffalo, which thrive better in drought than domestic animals. What a vindication that would be of the American Indian philosophy of living in harmony with nature rather than plundering its bounty!

The IIASA report paints a cheerful picture of agricultural prospects in the central region of the USSR, around Moscow. It estimates that a 1°C rise in temperature will increase grain production by 14 per cent. By investing 30 million roubles (around £30 million) in equipment, labour and storage facilities, to take advantage of the improved growing conditions, the USSR could increase output by another 10 per cent. The prediction confirms the optimism of Mikhail Budyko, the Soviet pioneer of global warming studies. Budyko argues that evidence from the previous periods of warmth 6,000, 125,000 and 3 to 4.5 million years ago suggests that global warming will be good for agriculture over the next half century. Problems will come only in the longer term. Higher temperatures, higher rainfall, an atmosphere richer in carbon dioxide, and an extension of the areas suitable for growing crops could increase harvests by about 50 per cent. However, Budyko admits that his conclusions are not universally accepted.

In Budyko's view, the global warming which has already occurred this century has brought a reduction of moisture in the USSR and other European countries. In the next century, increased warming should bring greater rainfall. As he explains:

The climate trend towards aridity in the US seems to
start later than in Europe and, by the year 2000, it
might appear to be more significant, although there are
not sufficient grounds to consider that the consequences would be disastrous. One might believe that
in the US the moisture conditions will start to improve
in the first quarter of the century.

Budyko's views are highly controversial and currently
attract little support from other climatologists either
within or outside of the Soviet Union.

In Japan, the IIASA study suggests that global warming may bring an embarrassing glut of agricultural produce. Japan's 4 million farmers already produce more
rice than its 123 million people can eat. By the year 2040, a
temperature rise of 3 to 5 °C will greatly increase the area
suitable for rice cultivation, leading to a tripling of stocks
within sixteen years. A similar picture emerges in Iceland, where the main agricultural activity is sheep and
cattle farming. A 4 °C temperature rise is likely to be
accompanied by a 15 per cent increase in precipitation,
and a growing season that starts forty-eight days earlier
than now. The hay harvest could increase by 66 per cent,
and cultivated grassland could support two and a half
times as many sheep. But pests and diseases, which are
now kept at bay by Iceland's chilly winters, will also
proliferate. Icelanders could take a tip from the barley-
growers and beef-fatteners of north-east Scotland, which
has a climate similar to that predicted for their country.

Preliminary studies in Britain suggest that a 3 °C temperature rise would reduce the yield from cereal crops
and increase the yield of root crops like potatoes and
sugar beet. Changes in rainfall, about which there is
great uncertainty, could worsen the picture. Less summer rain would reduce yields considerably, and more
rain, particularly in winter, could increase the loss of
fertilisers from the soil and make it more difficult for
farmers to cultivate their land in autumn. More at-

mospheric carbon dioxide would improve wheat yields only if there was no temperature increase as well. But for potatoes, a combination of more carbon dioxide, a 3 to 4°C temperature increase, and more summer rainfall would bring a bonanza increase in yields of between 50 and 75 per cent. The areas suitable for growing crops would extend northwards and into the uplands. In the south, more exotic crops could be introduced. But the major predictable impact of the greenhouse effect, says Simon Gourley, President of the National Farmers Union, 'will be its very unpredictability. We expect that the south-east will become drier and the north-west wetter but it is the high incidence of climatic extremes which may be most disruptive to farming.'

If global warming brings agricultural benefits to northern Europe, its agricultural effects could be disastrous in southern Europe, which lies close to the deserts of North Africa. Mediterranean Europe is already short of water, its soils are being eroded and invaded by sea salt, its river valleys are prone to flooding, and its forests are regularly destroyed by fires. The effects of climatic change in the area is one of the key elements in an EEC research programme which started in 1988.

The IIASA study is a document of limited usefulness Its estimates are not 'predictions of what will actually happen to agriculture in any particular place, but simply indications of how sensitive farming in various regions is to a changing climate,' says the study director, Martin Parry, of the University of Birmingham. Major uncertainties over regional and seasonal patterns of rainfall mean 'we simply do not have sufficient information, at this juncture, to indicate whether food supplies at national levels would tend to increase or decrease as a result of changing climate.'

The report makes detailed recommendations of how farmers could adapt to climate change by planting different crops and altering irrigation, crop rotation and other practices. In essence, it tells farmers and agriculture ministries: 'we do not know exactly what your problems are going to be, but these are the solutions you can apply when they appear.' It does, however, suggest that if the future means more droughts and storms the best strategy for farmers may be to adopt crops and methods which cope better with extremes rather than those which wring the highest possible yield out of the most favourable years.

In many temperate areas – the Nordic countries, northern Europe, central Russia and Japan – there may continue to be agricultural surpluses of the kind which have recently prompted action by the European Community to take land out of production, though this will depend heavily on the availability of moisture. The problem may be keeping farmers in business rather than keeping people in food. To say this, however, is to take a narrowly agricultural view of the situation. In conditions of changing climate, farmland may have a new future in growing trees to soak up carbon dioxide and crops which can be used to meet our energy needs on a more sustainable basis than coal, gas and oil. It remains to be seen, moreover, whether the food-growing capacity of these regions will be needed to feed the expanding populations of the developing world whose farmlands may come under even greater climatic stress.

Impacts in the tropics

Global warming will bring a relatively small increase in average temperature in the tropics, perhaps 1°C by the middle of next century (see fig. 2), but the effects on agriculture will be disproportionately large. They may also be felt sooner than in the higher latitudes. Farming in many parts of the tropics is a precarious activity, even more vulnerable to small variations in the weather than it is in other regions. Many parts of the zone are either arid or semi-arid: there is barely enough moisture for crops and natural vegetation to survive. A minor change in the distribution of rainfall throughout the year can make a significant difference to the success or failure of crops. These areas are already subject to frequent drought, with consequences which are now well known to us all. Other parts of the tropics depend on annual rainy seasons – monsoons. Here again, a change in the timing of the rains, or their volume, can have serious consequences. Flooding, already a problem in many places such as Bangladesh, may be exacerbated. Whether recent droughts and floods are harbingers of global warming is uncertain. Extreme climatic events throughout the tropics are linked to atmospheric and oceanic perturbations in the Pacific the cause of which has yet to be established (see 'The Troublesome Twins', pp. 62–63).

Studies by the IIASA team on the impacts in the tropics dealt mainly with short-term climatic variations. Their report concludes that the most significant effects may be substantial year-to-year variation in the yield and quality of crops and significant shifts in the boundaries of crop-growing areas. Crops grown in the tropics, they add, are highly sensitive to quite small changes in climate. Their dependability may be significantly altered by long-term climatic change. Once again they recommend that farmers consider changing their methods and choice of crops.

They also suggest that crop growing areas should be 'screened' to enable agricultural activities to be more closely matched to regional weather types. Combined with the improved use of new technologies, such as high-yielding crop varieties and consistently applied drought policies, this could reduce the intensity and extent of losses.

The question remains whether the generally impoverished developing countries of the tropics, with their fast-growing populations, can cope with such pressures and challenges. Clearly, global warming will

The Troublesome Twins

In the autumn of 1982, sea temperatures rose by 10°C off the coast of Peru. Plankton and anchovy failed to develop and the fishing industry collapsed. At the same time, there were enormously increased harvests of shrimp and scallop. In the following months, torrential rain fell in dry areas of California and Equador, forests caught fire in rainy New Guinea, and there were droughts in Australia, South Africa, India and Indonesia.

The Peruvians are used to such strange events, which happen every few years when a change of wind direction brings a surge of warm water across the Pacific. It begins just before Christmas, so they call it *El Niño* – the Child. Research in the last fifteen years has shown that it is linked with high global temperatures and extreme climatic events in various parts of the world. The cause is uncertain. Recent studies suggest it may be triggered by lava flows from undersea volcanoes.

The concentration of extreme climatic events in 1982–83 was due partly to *El Niño* and partly to the eruption of the El Chichon volcano in Mexico, which threw dust clouds up to 35km high, producing cooler weather in many regions. There was another *El Niño* in 1987.

Scientists have discovered a mirror image of *El Niño*, a pool of unusually cold water in the Pacific, which they call *La Niña* – the Girl. *La Niña* has the opposite effect to *El Niño*. Indian monsoons are drier when the Child appears, and wetter when his sister comes along. *La Niña* turned up in 1975 and global temperatures slumped in 1976. She appeared again in 1988, prompting some meteorologists to forecast a return to lower global temperatures in 1989.

The 'Troublesome Twins' play havoc with climate change forecasting. They are at least partially responsible for the run of record hot years and severe weather anomalies in the 1980s which could also signal the

onset of global warming. *El Niño* and the eruption of El
Chichon may also account for considerable variation
in ozone concentrations since 1982. But the Twins are
giving us a taste of what is to come, says Kenneth
Hare, chairman of Canada's Climate Programme Planning Board. 'We already get entire years that resemble
what will be normal five decades from now. We are in
effect allowed to rehearse the greenhouse warming,
and to examine its potential economic impact by inspection.'

worsen their existing problems. In chapter 4, we examine
these and other issues of particular relevance to the
developing world.

The squeeze on the forests

So to twice five miles of fertile ground;
With walls and towers was girdled round;
And there were gardens bright with sinuous rills;
Where blossomed many an incense-bearing tree;
And here were forests ancient as the hills,
Enfolding sunny spots of greenery.
                    Samuel Taylor Coleridge, *Kubla Khan*

Less than a ninth of the Earth's land surface is cultivated.
A similar amount is capable of being brought under
cultivation. About half the ice-free surface is grassland of
one sort or another, from alpine meadows to the vast
open spaces of the steppes and savannahs, which support herds of domesticated and wild animals. Some of
these areas have a light covering of trees and represent an
intermediate stage between open grassland and the
other great family of natural ecosystems, the forests.

Dense forest and open woodland still cover nearly a third of the land surface but both are rapidly disappearing in many parts of the world.

Evergreen coniferous forests are the natural eco-systems of Scandinavia and the northern parts of the USSR, China and North America. On high land, the pines and spruce reach further south into areas otherwise dominated by deciduous forests of oak, beech and other broad-leaved species. In the tropics are the evergreen broad-leaved rainforests, fringed by the seasonal leaf-shedding forests of the monsoon belt.

About a third of the natural forest cover has disappeared over the last 8,000 years or so, cleared by man for cultivation, settlements, fuel and timber. Early civilisations removed much of the forest cover of the Mediterranean region. More recently, large areas of North America have been felled. Forests account for 90 per cent of all the carbon contained in vegetation and their destruction is responsible for about half the extra carbon dioxide added to the atmosphere in the last two centuries. Although it has been overtaken as the main source by fossil fuel burning since the 1950s, forest loss is still a significant and possibly underestimated future source. The rapid destruction of tropical forests, particularly in Latin America and South-East Asia (which we examine in chapter 4), is a feature of the last thirty or forty years.

The boundaries of the different forest types and tree species have moved over the ages with the pendulum swings of global temperature. However, rapid global warming in the higher latitudes will leave less time for forests to migrate. The southern edges will become too hot, and perhaps too dry, for trees to survive. Northern latitudes will become suitable for forest growth but the trees may be unable to spread fast enough to keep pace. Forest belts therefore face a two-way squeeze unless they are helped out by extensive planting programmes.

Stephen Schneider maintains that the fastest rates of

forest movement that can be inferred from the end of the Ice Age is one kilometre per year, in response to temperature changes averaging some 1 to 20°C per thousand years. Predicted warming of 2 to 6°C over the next century is ten to sixty times faster. Studies by the US Environmental Protection Agency suggest that the southern boundary for hemlock and sugar maple, native species of the north-east states, will move northwards by 600 to 700km. Under the more extreme predictions of climatic change they would disappear almost entirely from the USA and would be confined, perhaps for centuries, to a narrow band across eastern Canada (see fig. 8).

As rising temperatures kill off mature trees and prevent the germination of seeds, forest decline may become visible in a matter of decades. Trees in the south-east and Great Lakes areas could begin to die back in thirty to eighty years. Some species, such as the balsam fir, found in North Minnesota, begin to shrivel with an average temperature increase of only 1°C. Michael Oppenheimer of the Environmental Defense Fund is concerned that global warming and increasing ultraviolet radiation will act simultaneously on forest ecosystems already stressed by other effects of human activity – acid rain, ozone, nitrogen over-fertilisation and soil contamination with poisonous metals. Oppenheimer says that these stresses may be interacting synergistically, each enhancing and accelerating the effects of the others. The EPA admits that the combined effect of such stresses 'cannot currently be determined'.

The economically important temperate forests of the middle latitudes will suffer most, says Jag Maini of the Canadian Forestry Service:

With billions of dollars being spent on tree planting programmes, in which the genetic characteristics of the seedlings are carefully matched with the current climate of the area in which they are planted, it is

The Impact of Global Warming

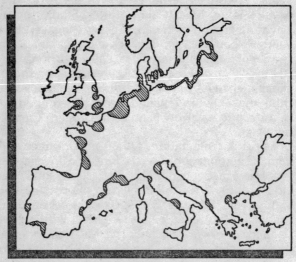

AREAS OF EUROPE VULNERABLE TO RISING SEA LEVEL

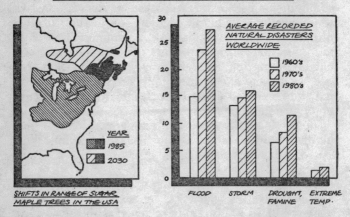

SHIFTS IN RANGE OF SUGAR
MAPLE TREES IN THE USA

AVERAGE RECORDED
NATURAL DISASTERS
WORLDWIDE

☐ 1960's
▨ 1970's
▨ 1980's

YEAR
1985
2030

FLOOD   STORM   DROUGHT,   EXTREME
                  FAMINE   TEMP.

Fig. 8

crucially important to discover how trees and seedlings will respond to climate change.

More important even than the growth behaviour of trees
may be the effects of further extreme weather. British
forestry plantations, mostly on thin, poor soils in exposed areas, are particularly vulnerable to gales. Hurricane force winds on 16 October 1987 in south-east
England destroyed fifteen million trees in a few hours.
Forest fires, relatively rare in Britain, were seen during
the dry summers of 1975 and 1976. They could become a
more regular hazard if, as predicted, much of southern
and eastern England develops a Mediterranean-type of
climate.

The Villach–Bellagio workshops concluded that forests are likely to be the main victims of climate change in
the mid-latitudes between 30 and 60 degrees North.
Their partial destruction by global warming may set up a
self-reinforcing feedback. More carbon dioxide will be
released and less will be absorbed, so atmospheric concentrations will grow even faster. The trees, other plants,
and soils of the northern forest belts contain as much
carbon as the atmosphere (see fig. 3). A warming rate of
0.8 to 1°C per decade, the fastest predicted for the mid-
latitudes, would start killing the forests by the year 2000.
The minimum predicted warming rate of around 0.06°C
per decade would delay destruction until 2100 and probably leave time for forests to migrate naturally.

The water comes and goes

For water to be a resource it must be at the time and place
required, in a suitable quantity and quality.

Jaromir Nemec, FAO, 1988.

Warmer temperatures will increase both the demand
for water and the amount of moisture evaporated from

soils, rivers, lakes and reservoirs. Changes in the pattern
of rain and snowfall may alter the flow of rivers and the
availability of underground water.

Jaromir Nemec of the UN Food and Agriculture Organisation (FAO) has made some alarming calculations
about the relationship between rainfall and the amount
of water that runs off the surface into rivers and streams.
A 25 per cent reduction in rainfall, he estimates, reduces
the runoff by about 80 per cent in dry temperate areas, 60
per cent in wet temperate areas, and 70 per cent in wet
tropical areas. The effect becomes even greater when the
effort needed to collect and store the water – to put it in
the right place at the right time – is taken into account. In
one case study, a 25 per cent reduction in rainfall required a 400 per cent increase in storage capacity to
maintain levels of supply. Impacts on this scale are far
greater than the variations in rainfall and run-off allowed
for in existing water management schemes.

Many countries are already chronically short of water.
In the developing world, where few people have access
to pure and plentiful water, 80 per cent of all illness is
attributed to unsafe and inadequate water supplies and
sanitation. Almost everywhere, underground sources,
rivers and lakes are becoming increasingly polluted.

Even in the US, which has a 'bountiful supply' according to the EPA, water is not always in the right place at
the right time or of the right quality. The EPA foresees
particular problems in California which depends on
plentiful rain and winter snow in the north to supply the
population centres and highly productive farmlands in
the south. With higher temperatures reducing the
amount of snow lying on high ground over winter, there
will be more run-off in the winter than the storage
reservoirs can hold, and not enough in the spring and
summer. If the river flows decrease, saltwater will
rapidly creep into the heavily farmed Sacramento–San
Joaquin river delta. Water supply is expected to fall by
7 to 16 per cent, while demand – for irrigation, to repel

salinity, and to meet the needs of a population expected to grow from 27 million to 36 million over the next two decades – increases.

Water supplies are already under strain in many parts of the USA. 'Farms and cities from Salt Lake City to San Diego,' says *Time* magazine, 'are literally drinking dry the Colorado River, which now peters out, exhausted and polluted, in the Mexican desert, miles short of the sea.' The huge Ogallala aquifer, which provides the Great Plains with underground water, is being overdrawn at the rate of 0.8 metres a year. Irrigation schemes have been abandoned in some areas.

The US shares with Canada no less than 18 per cent of the world's total volume of fresh water in the Great Lakes. There, global warming is predicted to cause two contradictory effects. The level of the lakes will fall, by perhaps 0.5 metres in Lake Superior and up to 2.5 metres in Lake Michigan. Ports and channels will have to be dredged to maintain the busy shipping lanes on the lakes, which have access, by the St Lawrence River, to the Atlantic. But the duration of winter ice will be reduced, by one to three months on Lake Superior and two to three months on Lake Erie, so ships will be able to ply the waters over a longer period. The smaller volume of water in the lakes is likely to exacerbate the already serious pollution problems.

The rising waves

The Sea washes all man's ills away.
Euripides, 484–406BC

We are an endangered nation.
President Maumoon Abdul
Gayoom of the Maldives, 1987.

President Gayoom's country, a collection of 1,196 low-lying atolls inhabited by 177,000 people, will virtually disappear under the waves of the Indian Ocean if sea levels rise by just one metre. The highest land in the islands is 2.5 metres above sea level and would be exposed to sea surges with a one metre rise. Other, larger nations are also at risk as global warming expands the Earth's surface waters and adds to their volume by melting glaciers. Four-fifths of Bangladesh, a nation of more than 112 million people, is flat delta land exposed to the rising waves along a front of some 650km. Much of Egypt's fertile land, much of its industry, and many of its people occupy the Nile Delta (see 'The Disappearing Delta', p. 71). Many of the world's biggest cities lie close to the sea on coasts and estuaries. Deltas and other low-lying coastal areas are among the most fertile farmlands in the world, while swampy coastal areas are among its richest natural ecosystems. Low-lying regions already suffer some $6–7 billion damage and 20,000 deaths annually from climate-related events.

Unlike temperature increases, sea-level rises will be relatively constant around the globe. Differences in the apparent increase will occur locally because some land surfaces are slowly sinking, and others rising, as the Earth's surface readjusts to a release of pressure following the end of the last Ice Age 18,000 years ago. For this reason, the general rise will be magnified on the south and east coasts of England and moderated on the northwest coast of Scotland.

Sea-level rise has been calculated at between 20 and 165cm in response to a doubling of greenhouse gases. If the West Antarctic ice sheet were to slip free from its small number of bedrock anchorage points and break up in a warmer Southern Ocean, sea levels could rise by as much as 5 to 6 metres, but it would probably take a couple of centuries of rising temperatures to make this happen.

The impact of sea-level rises on nature and society will be far from uniform. Some countries are more vulnerable than others. There seems little that the people of the Maldives can do except flee to whatever country is pre-

The disappearing delta

The Nile Delta, one of the cradles of civilisation, has been growing cotton for 3,800 years. Today its crops of rice, wheat and maize are the main source of food for Egypt's population, which is over 50 million and growing at the rate of a million every ten months. One-third of Egypt's fish harvest is caught in its freshwater lagoons. Many of the country's power-stations, oil refineries, chemical plants and textile factories are sited there, on land seldom more than 2 metres above sea level. Along a frontage of more than 200km, the delta is protected from the sea by narrow banks of sand carried out of Africa by the waters of the Nile. Since the Egyptians built the Aswan High Dam in 1964 to regulate the river, the replenishing flow of sand has ceased; in places the banks are crumbling into the Mediterranean at a rate of more than 100 metres a year.

Over the next hundred years, the climate models predict the Mediterranean will rise by between 20 and 140cm. Crops, factories, fisheries and the homes of millions of Egyptian people are at risk. Maps produced by UNEP's Global Resource Information Database (GRID) show that a 50cm sea rise would flood 1,754km sq of farmland and the homes of 3.3 million people. Alexandria, with a population of 3 million, and Port Said, with about 500,000 residents, would be all at sea. The Suez Canal would be shorter by nearly 30km.

A 100cm rise would carry the sea over 4,467km sq and affect 5.3 million people. Another 50cm would increase the toll to 5,674km sq and 6.2 million people: 19 per cent of the country's food growing area and 14 per cent of its population.

Four other important Mediterranean deltas face similar risks: the Ebro in Spain, the Rhone in France, the Po and Venice in Italy, and Lac Ichkeul in Tunisia.

pared to take them. Mainland countries like Egypt and
Bangladesh have the option of defending their territory
against the rising waves. Some countries, such as the
Netherlands, already have substantial land areas lying
below sea level. The technical feasibility of keeping out
the sea is not in question. Finding the resources to do so
is another matter. 'The level of funding needed may be
prohibitive for many countries, particularly in the Third
World,' says FAO's Jaromir Nemec. 'No one knows how
many of the rice-growing deltas and flood-plains in Asia,
or of the world's low lying cities, will be inundated over
the next century,' says the Worldwatch Institute. 'The
coastlines where protection might prove necessary in
decades to come could easily total thousands of kilometres', the cost may be measured in 'trillions of dollars'.

The speed at which the seas rise is crucial. A slow rise
will give people more time to assess the consequences
and decide whether to defend or abandon vulnerable
areas. A slow rise will be kinder too on natural coastal
habitats which, if given time, can adjust. If the change is
too fast, even the richest nations will be hard pressed to
defend their coastal lands, structures and settlements,
and natural systems will be overwhelmed. If time and
resources are available, there will be difficult choices to
make between defending farmland and settlements on
the one hand and allowing natural habitats to readjust on
the other.

Stephen Schneider, one of the first climatologists of
modern times to sound public warnings about the greenhouse effect, once re-drew the map of the USA on the
basis of a 4.5 to 7.6 metre sea-level rise. He found that half
of Florida and Louisiana would disappear, together with
a tenth or more of Virginia, Delaware, and New York,
displacing nearly 16 million people. On the more realistic
projection of a one metre rise, a string of cities and resorts
built on sandbanks on the eastern and Gulf coasts –
towns like Miami, Galveston and Atlantic City – would
be vulnerable to serious flooding during hurricanes. The

US Army Corps of Engineers, which is responsible for coastal defences and water-works, has put the cost of protecting beaches at $3 billion. In Latin America the beaches of Copacabana and Ipanema will disappear unless they are constantly replenished.

Permanent inundation beneath rising sea waters is only part of the problem. Coastal land which is not actually drowned may be invaded by saltwater, making agriculture impossible and spoiling underground water resources. Drains and sewers will be disrupted. River flooding will increase. Other effects of global warming will compound the problems. Storms will become more frequent and more severe. Storm surges – bodies of sea-water carried forward by high winds – will penetrate further inland. Wind, tide and wave patterns may change.

Coastal areas are in a constant state of flux. Cliffs and headlands are eroded by the tides; shingle beds, beaches and dunes are moved; deltas are created by the silt carried out of continents in rivers; marshes form and disappear. All these natural processes are liable to be disrupted, and to become less predictable.

Few nations have more experience of defending themselves against the sea than the Dutch. Their achievements in reclaiming land from the sea and defending their coasts from storms and sea-level rise show what can be done by those who have the will and the resources. According to Tom Goemans of the SIBAS Joint Institute for Policy Analysis in Delft, the first efforts can be traced back to the third century BC, when people built mounds to live on. The earliest written records describe the Coastal Dutch as 'water-men and mud-workers'. During the Roman period, the mounds were linked by elevated roads, and dykes were built along some of the rivers. By the thirteenth century the Dutch were building dykes not only to defend their land from flooding and erosion by the sea but also to extend their territory by creating polders, or protected areas of land, which they drained at

high tide with windmills. A long history of invasion by
storm surges culminated in the great storm of 1953 when
1,835 people were killed and 150,000 hectares were inundated.

The 4 metre surge experienced in 1953 was an event
calculated to occur once in 250 years. The Dutch decided
to play safe and rebuild their dykes to withstand a
5 metre surge, likely to occur once in 10,000 years.
Allowance was made for a sea-level rise of 25cm over the
next century. The construction programme has taken
over thirty years and will not be completed until 1990.
Goemens maintains that raising the dykes and dunes
and carrying out necessary work on water management
schemes, rivers and ports to cope with greater sea-level
rises would cost $2.2 billion for every 50cm step, spread
over twenty years. Coping with a 2 metre rise over a
century would cost somewhat less than 0.1 per cent of
the Dutch gross national product. If a 2 metre rise seems
likely, work will have to start almost at once, and in
collaboration with neighbouring countries. Monitoring
the sea-level change is extremely important in order to
detect the 'signal' as early as possible.

Coastal habitats and ecosystems

A key issue in society's response to rising sea levels is
how much land should be protected for the sake of its
existing economic uses and how much should be 'sacrificed' so that natural coastal habitats can re-establish
themselves. The conservation of wildlife – particularly
wildfowl, waders and sea-birds – is an important element
in the calculation. But it is not simply a question of people
versus nature.

Coastal wetlands are much under-appreciated sources
of wealth and well-being of the kind that seldom enters
into the calculations of accountants. They are, says
Edward Maltby, a British wetlands expert, 'essential life-

support systems' playing an important role in controlling water cycles, and helping clean up the environment by absorbing and neutralising pollutants. They also support economically significant national and local fisheries, and are important breeding grounds for many other fish species. The original home of rice, the oil palm, the sago palm and the mangrove forest (which yields timber, tannin, and many other products), wetlands are a biological gene bank second only to the tropical rain-forests. Important for recreation and tourism, they also have a further economic potential, now being explored, for 'aquaculture', in which fish and crops are reared together. Man's response to this bounteous ecosystem has been to systematically destroy it by conversion to conventional farmland and other 'improvement' schemes.

Burma and Malaysia have both lost more than half of their extensive wetlands in the last century. So has the USA, where a one metre sea-level rise will take 25 to 65 per cent of what remains, even if their natural migration is not blocked by the construction of coastal defences to protect settlements and farms. If all shorelines were protected, the losses would be 50 to 80 per cent. Louisiana, which has 40 per cent of the remaining US coastal wetlands – breeding ground for half the nation's shellfish – is already suffering large losses because the land is sinking at the rate of 15cm a century.

Coastal marshes, swamps, mud-flats, gently sloping shores, coral reefs and continental shelves are home to around 90 per cent of all marine species. Sea-level rise is not the only threat to their future. Changes in water temperature and salinity may affect the growth of coral. Fish and marine invertebrates, which cannot regulate their body heat, are also sensitive to changes in water temperature. More than just heat is involved, says Carleton Ray, a research professor at the University of Virginia. 'Warmer water contains less oxygen, and water temperature can influence the effect of pollutants on

marine creatures.' Many species may have to move, as
the climate changes, to friendlier habitats – or perish in
the attempt. The extinction of just a few species in any
habitat may have a knock-on effect on the abundance and
even the survival of other species which form part of the
ecosystem.

Big temperature increases expected in the polar regions may have a particularly drastic effect. Algae, on
which fish and ultimately birds and mammals depend for
food, grows on the underside of seasonal sea ice, says
Vera Alexander, director of the University of Alaska's
Institute of Marine Science. Seals breed on the icepack.
Warmer water and less ice are likely to spell extinction for
seals, walruses, and polar bears, and endanger bird and
marine species whose lifecycles take them from the polar
regions around the globe.

On land, as in the seas, species live in such close-
coupled relationships with one another and with their
environment that the effects of rapid climatic change may
be incalculable. On land, as on the coasts, human activity
and man-made structures compound the threat. Robert
Peters, a research scientist with the World Wide Fund for
Nature (WWF), explains:

> Under the present conditions of a human altered
> Earth, many species will be unable to make the necess-
> ary shifts because their routes will be blocked by roads,
> cities and 'agricultural deserts'.

Nearly all ecosystems will be affected to some degree,
whether by local temperature changes, alterations in
precipitation, soil and water chemistry, or sea-level rises.
At a 1988 WWF conference on the biological consequences of the global warming, Peters expanded on his
warning:

> Many species are already reduced in abundance, with
> some of their original ranges already used for human

activities. If some or all of their present ranges become unsuitable because of climate, and new habitat is not available to them, they may face extinction. Given substantial warming, many national park and other reserves would not preserve all the species now within them, because the habitats within the reserve boundaries would change in character.

One answer, he and many other ecologists believe, is to link nature reserves with networks of similarly protected migratory corridors.

The UK Nature Conservancy Council (NCC) has not yet developed its ideas for dealing with climate change. It faces, according to Brian O'Connor, the NCC's British Director, a 'revisionist period' as its focus shifts from safeguarding particular sites to considering ecological change:

One of the odd things about our country is that we are at the extreme ends of cold conditions in the north and warm conditions in the south. Many of our species are adapted to a particular and rather narrow climate range. I can imagine that many species that are at the northern end of their climatic range are going to extend northwards while those at the southern end are going to retreat from our shores. I think we could see quite significant changes.

Among the species at risk could be Britain's native frogs, toads and newts which are already under pressure because of the loss of breeding ponds. Warmer winters would help them but dryer summers could be a problem, says Mary Swan of Leicester Polytechnic, who is coordinating an NCC-funded survey of breeding sites. Waiting in the wings is an invader which could take over: the African clawed toad, the species used in laboratories for pregnancy testing. Escaped specimens are already

colonising Wales. They tolerate a wide range of climates;
and they eat the tadpoles of British species.

We know too little about the sensitivity of many
natural ecosystems to rapid climatic change to predict
what may happen. According to Norman Myers, a
British consultant on environment and development to
many national and international institutions, 'we are in
danger of underestimating the risks if we do not take
account of the combined effects of climate change and
other forces – land clearance, over-exploitation of natural
resources, and pollution – which are already endanger-
ing species and habitats. The consequences are likely to
be synergistic, that is, the total result of destructive forces
acting in conjunction with one another will be greater
than the sum of each effect taken on its own. Efforts to
understand the risks and to reduce the damage must be
made with all due urgency.'

Under the weather

What dreadful hot weather we have! It keeps me in a continual state of inelegance.
                        Jane Austen, *Letters*, 18 September 1796.

'Feeling under the weather, are we?' the doctor asks.
Human health and climate are indeed closely connected
(see fig. 8). Insect-borne diseases, a major cause of death
and illness in the tropics, are dependent on climatic
conditions. A strain of mosquito which killed tens of
thousands of people in Madagascar in 1988 may have
thrived due to a slight temperature rise. Extremes of heat
and cold, and other atmospheric features such as humidity, influence bodily functions and responses. Other
things being equal, people feel better and cope better in
the climate to which they are accustomed.

Impacts on Humans

DEPLETED OZONE LAYER
LARGE INCREASES IN SKIN CANCER

SEA LEVEL RISE
SALT CONTAMINATION OF DRINKING
WATER, SEWAGE SYSTEM FLOODED,
INCREASE IN WATER-BORNE
DISEASES, ABANDONED SETTLE-
MENTS

INCREASED DROUGHTS
AND OTHER CATASTROPHIC EVENTS,
CROP FAILURE, FOOD SHORTAGES
AND FAMINE

HIGHER HEAT AND HUMIDITY
INCREASE IN HEART ATTACKS, INSECT-
BORNE DISEASES AND VIOLENCE

Fig. 9

Extreme weather events such as heat waves take their toll
in additional deaths, particularly among the elderly and
the chronically ill, and can have other untoward effects,
such as increased discomfort and irritability leading to
carelessness, accidents and increased violence. Where
heat waves and drought lead to shortages of food and
water in circumstances compounded by poverty, the
impact on health and life are only too well known.
Norman Myers' observations about the synergistic effect
of climate change and existing stresses on natural ecosystems apply to mankind too.

Higher air temperatures and an increase in the amount
of ultraviolet radiation (see 'Danger – UV-B', pp. 81–83)
will accelerate chemical reactions in the atmosphere leading to the formation of secondary pollutants like ozone.
Ozone is highly toxic even at low concentrations, causing
immediate breathing problems – coughing, wheezing,
shortness of breath and chest pains – and in some cases,
longer-term lung damage and decreased resistance to
chest infections. In the US, where photo-chemical smog
is already a serious problem, the EPA estimates that a 4 °C
temperature increase in the San Francisco Bay area will
increase ozone concentrations by 20 per cent, without
allowing for any change in the emission of primary
pollutants.

Changes in wind speed and direction, rainfall patterns, cloud cover, atmospheric water vapour levels, and
global circulation patterns – all associated with climate
change – are liable to alter the way pollutants behave and
move around in the atmosphere. There is also evidence
that hot damp air may worsen the health effects of
breathing ozone-loaded air.

Episodes of elevated ozone concentration are com-
paratively rare in Britain, although far from unknown. A
London radio station now broadcasts details of air-
pollution levels, after ozone levels twice the World
Health Organisation recommended limits were measured in the autumn of 1988. Government scientists

Danger – UV-B

Ultraviolet radiation from the sun harms human and plant life by altering the structure and behaviour of proteins and nucleic acids in cells. The damage is done by radiation in the waveband of 290 to 320 nanometers, known as UV-B, which is partially filtered by stratospheric ozone. At the equator, up to 30 per cent of the UV-B entering the atmosphere reaches the Earth's surface. At the higher latitudes the proportion falls to 10 per cent.

UV-B causes sunburn and ages the skin. It also causes skin cancers. Two types of cancer are involved: localised skin cancers, which usually respond to treatment; and the general skin cancer melanoma, which is rarer but often fatal. These cancers are seldom found on negroid people because melanin, which gives them their dark colouring, is an effective filter of UV-B. Those at most risk are fair-skinned people with reddish hair, of Celtic origin, who freckle rather than tan when they sunbathe. Short periods of exposure to high intensity UV-B is particularly damaging. Robin Russell Jones, a London dermatologist maintains that:

> Girls who wear bikinis or sunbathe nude in early adult life carry a risk of developing melanoma on the trunk thirteen times greater than those who wear one piece swim-suits. Adolescents with fair skin, fair hair and numerous freckles carry a risk thirty-seven times greater than those with dark skin, dark hair and few freckles.

White-skinned people living an outdoor life in the tropics are particularly at risk. Non-malignant skin cancer is more than three times commoner than all other cancers combined among white Australians. By the age of seventy-five two out of three will have needed treatment. In Queensland, where the incidence of melanoma is ten times higher than in the UK, locally born white males stand a chance greater than 1 in 40 of developing the condition, and a 1 in 150 chance of dying from it. In England and Wales, melanoma kills about 1,000 people a year, and is now the fourth commonest cancer in women and the seventh commonest in men. It is a 'yuppie disease', seen most among indoor workers who holiday in the sun, observes Mark Elwood of Nottingham University Medical School.

CFCs and related gases have destroyed about 3 per cent of the global ozone concentration in the last twenty years, and about 6 per cent in British latitudes. The break-up of the 1987 Antarctic ozone hole produced a temporary 12 per cent depletion over Melbourne. Studies by the US Environmental Protection Agency suggest that a 1 per cent decrease in stratospheric ozone will increase the two types of localised skin cancer, basal cell and squamous cell, by 4 and 6 per cent respectively. A 5 per cent decrease will lead to cancer increases of 22 and 33 per cent. Worldwide, the Agency says, this would mean 70,000 new cases of non-melanoma skin cancer a year for a 1 per cent ozone depletion, and 360,000 new cases with a 5 per cent depletion. Melanomas are likely to increase by 1 to 2 per cent for every 1 per cent of ozone depletion, but the effect could be greater.

UV-B radiation also causes cataracts, one of the main forms of blindness. A 1 per cent ozone depletion may increase the number of cases by between 24,000 and 57,000 a year. Scientists now suspect that UV-B may pose an even greater threat to health by interfering with the ability of the body's immune system to fight off infections that enter by the skin. Among the conditions involved are Bilharzia, the parasite infection common in many parts of Africa, South America and China; leprosy, and herpes simplex. If further research bears out these fears, the global implications are much more serious than those for cancer and cataracts.

Rising levels of UV-B penetration pose a significant threat to phytoplankton, the single cell plants upon which almost all marine life depends directly or indirectly for sustenance. The larvae of many fish, including crabs, shrimp and anchovies, which live near the sea surface, are also threatened. In a study of anchovies, a 20 per cent increase in UV-B radiation over a fifteen-day period killed all the larvae in the top 10 metres of sea-water. Phytoplankton, which depend like other plants on photosynthesis, would reduce their growth rate if they moved into deeper water to avoid UV-B. The result could be a serious reduction in commercial fish harvests.

Tests on 200 land plant species showed that two-thirds reacted adversely to increased UV-B exposure, with peas, beans, squash, melons and cabbage among the most sensitive. Prolonged field tests on soy-beans – the fifth largest crop in the world – showed a 20 to 25 per cent reduction in yield from a 25 per cent depletion of ozone and the subsequent increase in UV-B, although some types appear less susceptible to damage than others.

predict that they will become an increasingly important problem as global warming takes effect.

This shrinking island – Britain and global warming

It is the year 2089, and the American and Russian tourists landing at Gatwick are pushing their way through terminal 5, mopping their brows, to join the trains for the Sussex Riviera. The old shingle beach at Brighton has largely disappeared under the waters of the English

Channel, but there are artificial sands in the air-conditioned pleasure domes (guaranteed UV-B proof). Restored after its partial destruction in the Great Hurricane of a century ago, the West Pier is now only one of half a dozen which reach out beneath the cloudless sky across the new sea-wall and promenade, giving access to marinas and sailboard pools. Dr Brighton's Bar is doing a roaring trade in chilled South Down Riesling. The more serious minded have departed for a spot of bird-watching in the new Suffolk wetlands. There are boat trips to the drowned town of Aldeburgh, once famous for its music festival, and the concrete mausoleums of the Sizewell nuclear power-stations.

Well, maybe . . . Predicting climate change in Britain is a tricky matter. Climate change models give a reasonably reliable indication of global temperature rises, but are 'absolutely hopeless' at showing how climate will change in the UK, admits Richard Warrick, a senior associate in the Climatic Research Unit at the University of East Anglia (UEA):

> A best guess might be that the UK would experience a warming that is comparable to the global average. But this is very rough guessing. We have no idea how precipitation will change and no idea how such effects would be distributed over the year.

The task is particularly difficult, because the climate is strongly influenced by Britain's position on the western edge of the a large continental landmass, facing a great ocean. The Gulf Stream, which flows across the Atlantic from the Gulf of Mexico, makes the climate damp and temperate, allows semi-tropical plants to grow on the west coast of Scotland, and keeps the Arctic coast of Norway free of winter ice (see fig. 10). The Atlantic will continue to influence Britain's climate, but in what way, climatologists cannot yet predict. Global warming is likely to alter both wind patterns and ocean currents. Irving Mintzer, of the World Resources Institute, suggests that the Gulf Stream could change route, leaving the British Isles and Iceland cooler and wetter at the same time as the continental landmasses grow warmer. Mick Kelly, from the UEA Climatic Research Unit, says Britain is still likely to be bathed in warmer ocean currents even if the Gulf Stream *does* move northwards, but admits that the results are unpredictable. Abrupt shifts and reverses in ocean current direction cannot be ruled out. A reversal in North Sea currents would bring extensive chemical pollution to Britain's shores from rivers like the Rhine and Weser.

Initial studies of climatic change in Britain, put in hand by the government before the Toronto Conference, assume an average temperature increase of 1.5 to 4.5°C, a sea-level rise of 20 to 165cm, and a change in rainfall of plus or minus 20 per cent. At the top of this temperature range, north-east Scotland would become as warm as south-east England is currently, and the south-east would have a climate similar to south-west France, with summer temperatures around 32°C (90°F). The maritime influence would continue particularly in western areas; southern, central and eastern areas would be drier and semi-Mediterranean; northern England and Scotland would be warmer and more temperate but not necessarily drier; winters could be warmer and wetter, and summers warmer and drier (see fig. 10).

The assumed rise in sea level puts areas at risk around the estuaries of the Thames, the Humber, the Tees, the Forth, the Tay, the Ribble, the Mersey, and the Bristol Channel, and low-lying coasts in Kent, Essex, East Anglia and Lancashire. More accurate topographical mapping is needed to show exactly where the waters would go if the coasts were not defended. The impact will be compounded by the gradual sinking of the southern and eastern coasts. Scientists are gathering evidence

Impacts on the United Kingdom

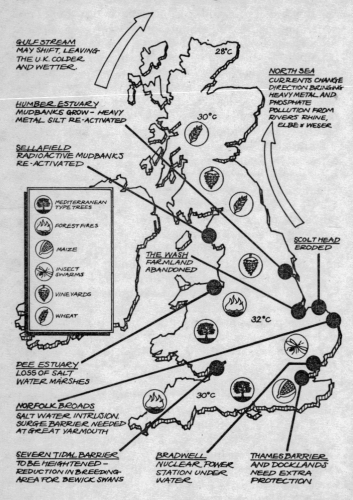

GULF STREAM
MAY SHIFT, LEAVING
THE U.K. COLDER
AND WETTER.

28°C

NORTH SEA
CURRENTS CHANGE
DIRECTION BRINGING
HEAVY METAL AND
PHOSPHATE
POLLUTION FROM
RIVERS RHINE,
ELBE & WESER

HUMBER ESTUARY
MUDBANKS GROW – HEAVY
METAL SILT RE-ACTIVATED

30°C

SELLAFIELD
RADIOACTIVE MUDBANKS
RE-ACTIVATED

MEDITERRANEAN
TYPE TREES

FOREST FIRES

MAIZE

INSECT
SWARMS

VINEYARDS

WHEAT

THE WASH
FARMLAND
ABANDONED

SCOLT HEAD
ERODED

32°C

DEE ESTUARY
LOSS OF SALT
WATER MARSHES

NORFOLK BROADS
SALT WATER INTRUSION.
SURGE BARRIER NEEDED
AT GREAT YARMOUTH

30°C

SEVERN TIDAL BARRIER
TO BE HEIGHTENED –
REDUCTION IN BREEDING
AREA FOR BEWICK SWANS

BRADWELL
NUCLEAR POWER
STATION UNDER
WATER.

THAMES BARRIER
AND DOCKLANDS
NEED EXTRA
PROTECTION

Fig. 10

of past sea-level changes and experimenting with new coastal defence techniques.

Sea coasts are never static. Tides and currents nibble away the land in one place and lay down new gravel and sand banks in another, while rivers deposit their silt in estuaries. Dungeness nuclear power-station in Kent has to be protected by regular 'nourishment' of the shingle banks on which it stands. Spurn Head on Humberside and Cromer Island in Norfolk were created by material carried down the coast by the currents. These processes may be affected by climate change which could, some suggest, reverse the direction of the North Sea currents. Changes in currents, tides and wave patterns may stir up the toxic metals which industrial pollution has deposited in the estuarial mud-flats and deliver the Rhine's huge outpourings of pollution to British shores.

Much of the British coast is protected by sea defences. Those on the East Anglian coast were renewed after the 1953 storm surge which breached the sea-walls in 1,200 places, drowned thousands of acres of land and killed more than 300 people. Those walls and breakwaters have now reached the end of their working life. They are being replaced in a £20 million a year programme aimed at protecting life, settlements, industrial structures and the drainage dykes of the Fens, the country's richest agricultural area. Elsewhere in East Anglia, thousands of hectares of low-lying farmland are deliberately being put at risk because the government and local authorities are not prepared to meet the costs of a full system of defences. In a report to the regional land drainage committee in January 1989, officials of the Anglian division of the National Rivers Authority noted the additional risk posed by global warming but concluded, 'the problems are essentially long term, circumstances may change and no clear scientific opinion has been formed.' Their recommendation was, essentially, to wait and see.

Government advisers have put the cost of building defences against a 20 to 165cm sea-level rise at £5–8 billion. They say that various options will have to be considered, including raising the existing sea-walls, building new ones further inland, building storm-surge barriers and estuary barrages, and even 'abandoning whole sections of the coast'.

The chances are that all these options will find a place in whatever strategy is eventually adopted. The 'soft' coasts most at risk from rising sea levels include the marshes, sands and mud-flats which are important wild-life habitats, where native and migrant birds feed in their hundreds of thousands. Without protection, a large proportion of them are at risk. An 80cm rise could lop 20 per cent off the mud-flats of Essex, while a rise of more than 1 metre could halve the mud-flats and virtually destroy the marshes. Left to its own devices, nature would re-establish many of these habitats on what is now dry land. We could lend a hand by building new sea-walls further inland. That might well be a prudent course, not only in the interests of conservation, but because marshes, mud-flats and dunes are natural protections against the sea. Such schemes are already being tried on several stretches of coast.

Most of Britain's natural coastal habitats are protected under national law, and some under international agreements. But many may be sacrificed to save towns, ports and industrial installations, according to Ted Hollis of University College London, who has been studying the issue for the WWF. Nearly three-quarters of the protected sites of special scientific interest are bounded by urban areas, leaving them little room to migrate naturally if property is protected. 'Sea-level rise is not the problem,' Hollis states, 'people are. It's their structures and responses that will determine what will happen to coastal and other wildlife habitats.'

Where the coasts are backed by low-lying agricultural land, conservation could have a better chance. According to Hollis, Titchwell Marsh in Norfolk is:

a shining example of the positive aspects of a sea-level rise for nature conservation. After the sea broke through in 1953, new salt marsh developed as farmland was abandoned. Little tern, bittern and Montagus Harrier were among the birds that moved on to the site. The Royal Society for the Protection of Birds bought it in 1973 and created a whole series of coastal wetland habitats in this 'disaster area'. The number of breeding bird species increased from thirty-nine to sixty-one, and there were 69,000 visitors in 1986. It can therefore be argued that nature conservation will benefit when the environment 'bites back' with sea-level rises.

The National Farmers Union is worried that abandoment of land is the option which will appeal most to the authorities. With intensive agricultural methods and constantly improved yields from both crops and livestock, farmers are producing more and more food on less and less land. This comes at a price, however, in the form of nitrate pollution in water supplies and damage to soils. If full advantage is taken of biotechnology, says John North, a former government agricultural adviser, now in the Department of Land Economy at Cambridge University, Britain could by early next century produce all the home-grown food it needs on little more than a quarter of its farmland. Economic, social and environmental considerations make such an extreme outcome unlikely. However, it is clear that British farming could adjust to the pressures of climate change and still occupy less land than it does today. John North is no advocate of abandoning land to the waves. All the land we have should be saved, he argues, for growing cereal and other crops – not as food, but as sources of fuel and chemical feedstock, replacing oil and gas. About 95 per cent of the goods made from these petrochemicals could be made from cereals by early next century.

The British climate has changed perceptibly this century. Records show us that the prevailing North Atlantic westerly winds were stronger, and the winters milder between 1900 and 1930. For the past twenty years or so it has been cooler over much of north and east Europe, in spite of the warmer average global temperatures. Whatever the long-term effects of global warming, Britain like other areas is in for more variable weather over the coming decades.

To say that the British climate is variable is to state the obvious, but the consequences of that variability are not always so evident, except to those businesses it effects. A study by the Atmospheric Impacts Research Group at Birmingham University shows that the annual value of British agriculture can vary by almost a third due to the weather. The turnover of the transport and communications industry can vary by about a fifth. The everyday vagaries of the weather cost British insurance companies some £800 million a year in property claims. Extreme events add well over £1,000 million to the bill every decade. Just one brief event, the storm of 16 October 1987, cost the insurers in Britain and France well over £1,500 million. With a greater frequency of extreme weather events, the insurance bill is set to rise, perhaps substantially, and the oscillations of industrial turnover may become even greater.

Increased severity and unpredictability in the weather poses another problem. Civil engineers, water-supply planners, and others responsible for taking long-term design, management and investment decisions, rely on past weather records, and especially the frequency of extreme events, to tell them how much allowance they must build in for floods, droughts, and high winds. Since no one can say how much more frequent and how much more severe the extreme events are likely to be in the future, the planners are in a quandary. Do they press on regardless using past experience, and take out more insurance? (But how do the insurers assess the risks?) Or do they build in a bigger margin of error and hope that they have not over-compensated? Time and research may provide some clearer answers. But by then it may be too late.

4

THE SECOND FRONT

The simple, and terrible, truth is that poverty and environment are inextricably linked in a chain of cause and effect. Problems of environment cannot be tackled in isolation from those national and international economic factors that perpetuate large scale poverty.

Shridath Ramphal,
Commonwealth Secretary-General,
February 1989.

1988 was an extraordinary year. The elements made war and men made peace. While droughts, floods, hurricanes and earthquakes took their toll of death and misery, the superpowers agreed to scrap part of their nuclear arsenals. The USSR withdrew from Afghanistan, announced plans to cut its forces in Europe, and paved the way for a similar retrenchment on its eastern borders with China. Iraq and Iran agreed a cease-fire. Palestine recognised Israel's right to existence. India and Pakistan started to heal old wounds.

It would be overstating the case to suggest that a few warning shots from nature were wholly responsible for the change of heart. Nevertheless, it was explicitly recognised by many world leaders in 1988 that the turbulence in the atmosphere and the calm in the conference chamber were something more than coincidence. The connection was traced by Shridath Ramphal, the Commonwealth Secretary-General, early in 1989. He was speaking in the first of a series of Cambridge Lectures on the report *Our Common Future* by the World Commission on Environment and Development, commonly called the Brundtland Report after its chair Gro Harlem Brundtland, Prime Minister of Norway:

> The interrelated issues of environment and development now vie with nuclear disarmament as the dominant issue of our time. Politicians – from Mr Gorbachev to Mrs Thatcher – and financiers – from the President of the World Bank to environmentally 'clean' unit trust managers – advertise their 'green' credentials. But, perhaps more important, was the way in which a succession of disasters all over the world triggered intellectual awareness about the possibility of some underlying pattern of causality, and aroused those emotions of fear and anger that are often the mainspring of political action.

If 1988 was the year in which the world's senior politicians first publicly acknowledged that human abuse of the natural environment had finally gone too far to be ignored, one day in particular – Tuesday 27 September – brought this greening of political consciousness into sharp focus with two remarkable and unexpected speeches.

In London, British Prime Minister Margaret Thatcher addressed the annual dinner of the Royal Society. As was fitting, she reminded Britain's most eminent scientists of the achievements of science. Medicine had saved millions upon millions of lives – and brought the global population to 5 billion. Agricultural research had found the means of feeding the 5 billion – and left a legacy of nitrate pollution and rising methane emissions. Engineering had given us the capacity and need to exploit fossil fuels – and brought a vast increase in carbon dioxide.

The Brundtland Report

The World Commission on Environment and Development was set up by the United Nations in 1983 to 'propose long-term environmental strategies for achieving sustainable development by the year 2000 and beyond'. Chaired by the Prime Minister of Norway, Gro Harlem Brundtland, its report, *Our Common Future* was published in London in 1987, and is often known as the 'Brundtland Report'.

The Commission defined 'sustainable development' as ensuring that humanity 'meets the needs of the present without compromising the ability of future generations to meet their own needs'. It is not a cost-free approach. Sustainable global development 'requires that those who are more affluent adopt lifestyles within the planet's ecological means.' It is 'a process of change in which the exploitation of resources, the direction of investments, the orientation of technological development, and institutional changes are made consistent with future as well as present needs. We do not pretend that the process is easy or straightforward.'

The Commission concluded that the needs of a growing world population could only be met by making more efficient use of energy resources. On global warming, it called for an immediate start on improved monitoring and assessment, increased research, the development of internationally agreed policies for reducing greenhouse gases, and the adoption of strategies to minimise damage and cope with change.

Gro Harlem Brundtland told the Toronto conference in 1988, 'It is established beyond any doubt that we *will* experience a global change in climate.' She called for an international action plan, including immediate discussions on the feasibility of adopting regional strategies for stabilising and reducing energy consumption:

If we are serious in our attempts we must be prepared to tackle the myth that energy consumption
must be allowed to grow unchecked. In Norway, we
are now aiming at a stabilisation in energy consumption by the year 2000.

A change in Norwegian production and consumption patterns will only contribute marginally
to solving the global problem. Presently, developing countries must be allowed time for adaptation
and the chance to increase their consumption. Industrialised countries have a special obligation. We
must be the first to change.

The Prime Minister continued:

For generations we have assumed that the efforts of
mankind would leave the fundamental equilibrium of
the world's systems and atmosphere stable. But it is
possible that with all these enormous changes concentrated into such a short period of time we have unwittingly begun a massive experiment with the system of
this planet itself.

Three changes in atmospheric chemistry had become
familiar subjects of concern: the increase in greenhouse
gases, the discovery of the ozone hole, and acid depo-
sition.

In studying the systems of the Earth and its atmosphere we have no laboratory in which to carry out
controlled experiments. We have to rely on observations of natural systems. We need to identify particular
areas of research which will help to establish cause and
effect. We need to consider in more detail the likely
effects of change within precise timescales. And to
consider the wider implications for policy – for energy
production, for fuel efficiency, for reforestation.

The government espoused the Brundtland concept of
sustainable economic development.

> Stable prosperity can be achieved throughout the
> world, provided the environment is nurtured and
> safeguarded. Protecting this balance of nature is there-
> fore one of the great challenges of the late twentieth
> century.

In New York, the Soviet Foreign Minister Eduard Shevardnadze told the General Assembly of the United
Nations that the world community had reached a watershed in 1988:

> The growing physical destruction of our planet is the
> verdict against the existing divisions of the world.
> Perhaps for the first time we have seen the stark reality
> of the threat to our environment – a second front fast
> approaching and gaining an urgency equal to that of
> the nuclear-and-space threat. For the first time we
> have clearly realised that, in the absence of any global
> control, man's so-called peaceful constructive activity
> is turning into a global aggression against the very
> foundations of life on Earth. For the first time we have
> understood clearly what we just guessed: that the
> traditional view of national and universal security
> based primarily on military means of defence is now
> totally obsolete and must be urgently revised.
>   Faced with the threat of environmental catastrophe,
> the dividing lines of the bipolar ideological world are
> receding. The biosphere recognises no divisions into
> blocs, alliances or systems. All share the same climatic
> system and no one is in a position to build his own
> isolated and independent line of environmental de-
> fence . . . We need resources to save our planet, in-
> stead of destroying it. I think the world community has
> such resources. But they have to be supplemented by
> the will and readiness to act, and secondly, by an

effective mechanism for international ecological co-
operation . . . We have too little [time] and problems
are piling up faster than they are solved.

Under the mantle

When the leopard changes its spots, scepticism is the
order of the day. Margaret Thatcher's speech astonished
Britain. The Prime Minister, committed to economic
growth and the relaxation of public restraints on private
enterprise, had hitherto been manifestly impatient of
environmental concerns. Now, suddenly, 'the greening
of Maggie' was on everybody's lips as countless newspaper articles and chat shows attempted to fathom her
motives. The *Financial Times* columnist Joe Rogaly
dubbed her 'Mrs Greenmantle' after John Buchan's mys-
terious undercover hero.

An obvious explanation for Thatcher's 'greening' was
a growing tide of public concern about the environment,
measurable in successive opinion polls which showed
electors, not only worried about pollution and environmental degradation, but also prepared to pay more in
taxes and prices to remedy these defects. Opposition
parties were responding more positively to these issues,
which were highlighted by the enormous media interest
in August 1988 in the North Sea seal deaths and the
attempts to land a cargo of Italian toxic waste in Britain. It
was time for the Prime Minister to reassert her authority.

Another more sceptical view was that concern over the
greenhouse effect gave the Prime Minister an excuse to
promote nuclear power as a 'clean', carbon dioxide-free
source of energy. The public inquiry into plans for
Britain's second pressurised water reactor at Hinkley
Point was about to start and the nuclear industry, after
Chernobyl and with the electricity industry about to be
privatised, was facing some difficulty in making its case.
'This was a conversion,' said Jonathon Porritt, Director of

Friends of the Earth, 'not on the road to Damascus but to
Hinkley Point' (see chapter 6).

The most likely explanation, however, is that Thatcher
was growing worried about being pushed aside on the
international stage, where concern about global warming
was, and is, increasingly a central concern and topic at
summit meetings. Britain had not been totally silent in
international gatherings. The UK ambassador to the UN,
Sir Crispin Tickell, is a notable exponent of global warming concerns. But his voice was not being heard. The
government sent a message of commitment to tackling
the problem to the Toronto Conference in June 1988,
carrying the signature of Lord Caithness, Minister for the
Environment, Countryside and Water.

The statement stressed the uncertainties of climate
change, Britain's participation in the World Climate Programme, and its determination 'that if necessary we
should do more.' It continued:

> If we are to seek a preventative approach to environ-
> mental problems, I believe that we need to set our-
> selves a timetable of response which paces the likely
> development in our understanding over the next ten to
> fifteen years. In the next few years I see an especially
> vital role for research. [Even before the science is fully
> established] cost-effective measures to promote en-
> ergy efficiency must be encouraged and emphasis
> must be placed on realistic energy pricing, reflecting
> the true cost – including any environmental cost – to
> the customer.

Caithness was unable to attend the conference and the
statement received no publicity. For the Toronto Conference in June 1988 the government also produced a first
assessment of the likely impacts of global warming on
Britain (see chapter 3) which was briefly reported at the
time in *The Guardian*. There was also a brief reference to
global warming in the government's response to the

Brundtland Report (see 'The Brundtland Report', pp. 94–95), published the following month.

Only one positive policy initiative emerged in subsequent months: a commitment to strengthening the Montreal Protocol on reducing the emissions of CFCs. The Environment Secretary Nicholas Ridley and other ministers repeatedly stressed that this was the 'most important and immediate thing' the world could do to counter the greenhouse effect. It is also a relatively cheap option. But the British government notoriously proceeds by stealth. Its view of the 'immediate and optimum policy', set out in the Caithness message and by the Environment Department's chief scientist David Fisk in an address to the Royal Institute of International Affairs (RIIA) in October (both unpublicised at the time of delivery) is this:

- [ ] the wide ratification and strengthening of the Montreal Protocol;
- [ ] proper economic pricing of fuels on the world market;
- [ ] improved energy efficiency worldwide;
- [ ] better land use practices on the global scale;
- [ ] international effort to resolve major scientific uncertainties before the end of the next decade.

All these areas are presumably being dealt with in the reviews of departmental policy which Mrs Thatcher set in motion, and which she and other ministers are periodically examining in one of those Cabinet committees whose existence is only begrudgingly acknowledged and whose results will no doubt emerge in the fullness of time, probably in the form of a tactical leak. There seems little doubt, however, that the departmental dinosaurs are being slowly turned about.

Eduard Shevardnadze's speech was even more startling and, taken at face value, contained more substance. He proposed – to some alarm in the West – that the UN Environment Programme should be transformed into 'an environmental council capable of taking effective decisions to ensure ecological security'. He also proposed a series of emergency meetings under the auspices of the UN to co-ordinate efforts in the field of ecological security: a consultative meeting of experts in 1989 to discuss the health of the Earth; a summit meeting of the leaders of fifteen to twenty states 'representing all continents and the non-aligned movement' in 1990; and a second UN international conference on the environment 'to be held, as planned, in 1992 or even earlier, but in any event at the summit level.' This was less surprising, since UNEP was mapping out such a programme and world leaders were falling over one another to take the credit for bringing it into being.

President Bush, in an election campaign speech in August 1988, made a similar pledge:

> Those who think we're powerless to do anything about the greenhouse effect are forgetting about the White House effect. As President I intend to do something about it. In my first year in office I will convene a global conference on the environment at the White House. I will include the Soviets, the Chinese, the developing world as well as the developed. All nations will be welcome – and indeed, all nations will be needed.

Shevardnadze's speech was fleshed out by Mikhail Gorbachev at the UN in December 1988 when he explicitly linked his unexpected announcement of cuts in Soviet conventional forces with environmental and economic issues:

> The growth of the world economy reveals the contradictions and limits inherent in traditional type industrialisation. Its further extension and intensification spells environmental catastrophe. International economic security is inconceivable unless related not only to disarmament but also to the elimination of the threat

Greenhouse glasnost

The US and USSR have been working together for several years on global warming problems. In 1985, Reagan and Gorbachev signed an agreement to cooperate in the preservation of the environment. At their December 1987 summit they approved a bilateral initiative to 'pursue joint studies in global climate and environmental change.'

In February 1988, the joint commission directing the work agreed on a bilateral study of possible response strategies to climate change, including an examination of biological, economic and social impacts and a major exchange of data on the agriculture and water resource impacts, plus the identification of measures to reduce levels of greenhouse gas emissions. It also agreed to begin climate studies in the Arctic and measurement of ozone and methane changes in the Arctic and Antarctic. In May, discussions began on using space technology to study climate change. The two leaders renewed their promises of cooperation at the Moscow Summit in 1988. Reagan had the backing of a Senate Environment Committee who said:

> The United States and the Soviet Union are the world's two largest contributors of carbon dioxide. Together we account for almost one-half the global total. For these reasons, the United States and the Soviet Union must take positions of global leadership on this matter.

The Rocky Mountain Institute (RMI) in Colorado, a leading centre for energy saving research, has an exchange programme with Soviet scientists, dubbed 'citizen techno-diplomacy'. The Institute's first visit to the USSR generated a forty-five minute Russian TV film the *Energy Efficiency Revolution – A Key to Perestroika*. A Soviet delegation to the RMI in August 1988 included Yevgeni Velikhov, vice-president of the

Soviet Academy of Sciences and one of Gorbachev's most senior policy advisers.

During a series of meetings in Moscow in May 1988, Velikhov signed an agreement with the US Natural Resources Defense Council (NRDC), a non-governmental organisation, for a programme of 'anti-greenhouse' energy efficiency demonstration projects.

Greenpeace International meanwhile have initiated their own 'ecological *glasnost*', by agreeing a major record deal with the Soviets. Artists like Sting, U2, Dire Straits and Terence Trent D'Arby have donated songs for the disc, whose first pressing will be 4 million copies. Profits will go equally to the International Foundation for Survival and the Development of Humanity, which is headed by Velikhov, and new Greenpeace offices throughout the Eastern Bloc.

to the world's environment. In a number of regions, the state of the environment is simply frightening.

One of those regions is the Soviet Union, as Michael Heseltine, a former British Secretary of State for, successively, the Environment and Defence, was quick to point out in a speech to the Royal Institute for International Affairs (RIIA) in November 1988. The Soviet leaders had every reason to be concerned about the environment, he said. 'Their hopelessly inefficient economies do more than most to squander its scarce resources. And their out-of-date industrial process characterises the worst environmental legacies.'

☐ Chernobyl had left 33 dead, 2,000 more to die and 150,000 evacuated;
☐ in rice fields near Rostov, excessive pesticide use had

led to a 27 per cent increase in cancer over five years and birth defects up by 55 to 60 per cent;

☐ two West Siberian rivers, the Ob and the Irtysh, had twenty times the permitted level of oil products, and the Caspian Sea nine times the permitted level of phenol;

☐ a 13 metre drop in water level had converted the Aral Sea into the Aral pond.

The USSR, Heseltine argued, would use the need to clean up its environment and join in global warming studies to plead for resource and technology transfers from the West which would serve also to improve its military capacity. And the Soviet leadership, he warned, would exploit the growing environmental movement in the West as it had earlier exploited the peace movement:

> What we are seeing here is a well thought out, carefully crafted attempt to hijack the environmental agenda partly for ulterior motives. You do not have to be a cold warrior to recognise that a new era has been opened in which to fight some of the old battles. [Gorbachev] has spotted that the West has a long and vulnerable flank exposed on its environmental record. The message of the rise of the Green parties in Western Europe is clear. The environmental record of governments on both sides of the Atlantic, even when seen through the eyes of their own supporters, no longer matches the expectations of their democracies. Western leadership is now under threat for its perceived failure to respond early enough and adequately enough to the developing ecological crisis. We are witnessing the birth of Green geopolitics. We must be sure we are well prepared.

According to Heseltine, the Soviet Union had signalled its intention to make international environmental conferences into ideological battlegrounds.

Others take a more supportive view of the Soviet need to modernise its economy. Lester Brown and Edward Wolf wrote in the Worldwatch Institute's 1988 *State of the World* report:

> As market reforms penetrate the ossified Soviet economy, energy efficiency will climb sharply, eliminating some of the extraordinary waste associated with its centralised planning and management. The vast potential these reforms hold for reducing carbon emissions gives the entire world a stake in their success.

David Cope, of the UK Centre for Economic and Environmental Development, commented:

> If military tensions can be reduced by appropriate disarmament strategies, then the displacement of ideological conflict into competition into how '*green*' are different ideologies must be of benefit to the environment – a global-wide embodiment of competitive marketing. It is even possible that 'sustainability' concepts might lead to convergence between some aspects of the world's two dominant economic and political systems.

The Soviet bloc's industrial, economic and environmental problems are manifest, and in the era of *glasnost* it is making little attempt to hide them or to stifle internal public pressure for improvement. Already one of the leading producers of fossil-fuel derived carbon dioxide, its electricity consumption has been growing rapidly, and yet it has had to impose restrictions in Romania, Bulgaria, Poland and East Germany. In just one week, BBC monitoring of East European broadcasts revealed that:

☐ work had halted on two new nuclear power-stations, and the management of an existing one was under review;

☐ one major hydroelectric project had been abandoned and another severely cut back on environmental grounds;
☐ there were complaints about pollution from a shale oil-fired power-station under construction in Estonia.

The USSR boasts of a £53 billion budget for environmental protection between 1981 and 1986, claims that air pollution has been cut by a third, and effluent to rivers, lakes and seas by a third. Nevertheless, air-pollution problems abound and a number of industrial plants have been obliged to suspend operations, Judith Perera reported in the *New Scientist*.

In July 1988, the Estonian Communist Party authorised the formation of a Green movement, aimed at 'averting ecological catastrophe', pledged to campaign against nuclear power and for an alternative lifestyle. Fyodor Morgun, head of the recently formed USSR environmental protection committee told the party conference, also in July, that a 'grave ecological situation has emerged because of the ill-considered drive to build gigantic plants.' The natural fertility of soils was declining, topsoil was decreasing, forests were in an unsatisfactory condition, the air over industrial centres was polluted, and the quality of river water was worsening almost everywhere.

The roots of change

The greening of international relations is not some altruistic conversion to a belief in the sanctity of the natural environment. It is not just a recognition of the scientific evidence that natural systems have been grossly abused, although that has played a crucial part. It is certainly a response to growing internal political pressures. But one aspect in particular is in the forefront of world leaders' minds. It is summed up by the term **environmental security**.

Environmental, or ecological, security – the words tend to be used interchangeably – means something rather more than protection of natural resources and habitats. It encompasses the idea that environmental degradation and the increasing scarcity of what were once regarded as the free goods of bountiful nature – fresh water, productive land, mineral and fuel resources – are a cause of human conflict, and one which is becoming more acute with the onset of climatic change.

According to Norman Myers, an adviser to several UN agencies:

> If a nation's environmental foundations are depleted, its economy will steadily decline, its social fabric deteriorate, and its political structure become destabilised. The outcome is all too likely to be conflict, whether conflict in the form of disorder and insurrection within the nation, or tensions and hostilities with other nations. We can surely expect that this new scope for conflict will expand as increased numbers of people seek to sustain themselves from declining resource stocks.

Environmental factors, Myers argues, have helped foster food riots in the Philippines, Bangladesh, Egypt, Tunisia, Morocco, Zambia, Colombia, Brazil, Bolivia, Haiti and the Dominican Republic; brought Britain and Iceland to the brink of a Cod War; and triggered conflict in Ethiopia.

> The superpowers can hardly consider that the world as a whole will remain stable and secure if the Third World does not leave behind its absolute poverty, a good share of which derives from environmental degradation.

Our Common Future made the same point:

> The developing and widening environmental crisis presents a threat to national security – and even survival – that may be greater than well-armed, ill-disposed neighbours and unfriendly alliances. Already in parts of Latin America, Asia, the Middle East, and Africa, environmental decline is becoming a source of political unrest and international tension. The recent destruction of much of Africa's dryland agriculture was more severe than if an invading army had pursued a scorched-earth policy. Yet most of the affected governments still spend far more to protect their people from invading armies than from the invading desert.

The report went on to identify global warming as the most worrying of environmental threats to security:

> Climatic change would quite probably be unequal in its effects, disrupting agricultural systems in areas that provide a large proportion of the world's cereal harvests and perhaps triggering mass population movements in areas where hunger is already endemic. Sea levels may rise during the first half of the next century enough to radically change the boundaries between coastal nations and to change the shapes and strategic importance of international waterways – effects both likely to increase international tensions. The climatic and sea-level changes are also likely to disrupt the breeding grounds of economically important fish species. Slowing, or adapting to, global warming is becoming an essential task to reduce the risks of conflict.

Global warming is not something which is happening to a stable and secure world. It is an additional burden on a world already racked with shortages, insecurity and tension; a world whose exploding population is outstripping, if not the planet's capacity to provide, then society's ability to deliver, the necessities of life. It will not just destabilise the world's weather; it will also further destabilise an already grossly unbalanced human society. So let us look in more detail at where the heat will burn most fiercely.

The exploding world

June 5 each year is celebrated as World Environment Day. In 1987 it was chosen by the UN as the symbolic moment when the global population reached 5 billion. That represented a five-fold increase in less than two centuries. It took a century or more for the population to double from 1 to 2 billion: the date is placed sometime between 1918 and 1927. By 1960 the number was 3 billion, by 1974 it was 4 billion. The latest estimates based on revised fertility rates in India and China suggest that the 6 billion mark will be reached by 1998, a year earlier than previously expected. The growth curve may flatten out towards the middle of the next century but the figure is expected to reach 10 billion before stability is achieved in about a hundred years time (see fig. 11).

Three out of every four world citizens live in the less developed countries of the Third World, and an increasing proportion of them are huddled in and around its rapidly expanding cities. In 1985, Mexico City was the second largest metropolitan area in the world, with 17.3 million people. By the end of the century it is expected to overtake Tokyo (1985 population 18.8 million, 2000 projected figure 20.2 million) to top the league at 25.8 million. Sao Paulo had 15.9 million people in 1985 and is heading for 24 million; Calcutta had 11 million and is heading for 16.5 million. Cairo, to come down the UN-compiled ladder a little, had 7.7 million heading for 11.1 million, although by some estimates it has already

Pressure on the Planet

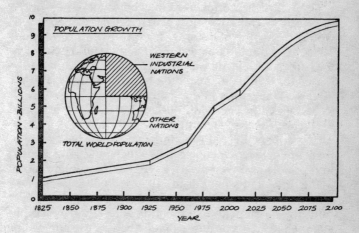

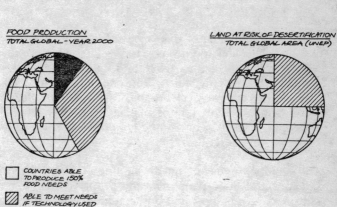

Fig. 11

passed that score. Dacca, the capital of Bangladesh, with a mere 4.9 million in 1985, is expected to reach 11.2 million by the year 2000.

Though their central areas may be as modern as any, these are not cities that most of us in the developed world would recognise. We know, from the work of Mother Theresa, about the people who live their whole lives on the streets of Calcutta and Bombay. In Cairo, an estimated three million make their homes in the city's cemeteries. 'You cannot find a map of all Cairo these days,' Robert Fisk, *The Times*' Middle East Correspondent wrote in 1986. 'The city is simply growing too fast for the cartographers. Even the modern maps show fields where there are now tenements, canals which have long turned into open sewers, cemeteries which now contain more living inhabitants than dead.'

In Rio de Janeiro, two-thirds of the city's shanty towns, housing an estimated 3 million people, are perched on steep slopes. According to Nefis Sadik, executive director of the UN Population Fund (UNPF):

Squatter settlements strip away the vegetation that anchors the soil and protects the watershed. With the bare soil exposed to wind and rain, the hillsides quickly become unstable. At the end of February 1988, while the city was celebrating Carnival, a mudslide following torrential rains claimed the lives of 275 people – mostly women and children – and left over 20,000 homeless.

The shanty towns of the developing world, built on hillsides and flood-plains, lacking proper roads, water supplies, sewers, hospitals and emergency services, are particularly vulnerable to the weather-related natural disasters – floods, hurricanes, and landslides – which will become more frequent as the planet becomes warmer. Some, like Dacca, are within the potential range of rising sea levels (see 'Living on the Margin', pp. 112–113). But the idea of planning ahead to prevent disasters or limit the scale of injury, death and disaster 'is hardly ever considered,' says Jorge Hardoy, director of the International Institute for Environment and Development's (IIED) Latin American office.

The first waves of shanty-town dwellers were probably driven by the hope of work and the dream of urban prosperity. Today the newcomers are increasingly recognised as 'environmental refugees', fleeing famine and land scorched by drought or so overexploited that it can no longer yield even the most meagre living. In 1984–85, an estimated 10 million Africans left their homes for city shanty towns and refugee camps, many crossing national boundaries. In 1986, at least a fifth of the population of the Ivory Coast were unofficial immigrants. The figures are not surprising, says Norman Myers, 'in a region where, in 1985, 150 million people faced food shortages, and where 30 million suffered from famine; the results not so much of recent drought as of long-standing environmental degradation compounded by ultra-rapid population growth and faulty development policies.'

Almost one million Haitians, a sixth of the population, have fled from an island once covered with trees, now almost totally denuded by fuelwood cutting. 'Erosion is occurring at such a rate that the peasant farmers believe that stones actually grow in their fields,' says Michael Andrews, a senior producer with the BBC Natural History Unit. 'During the rains, bulldozers have to clear the streets of Port-au-Prince of topsoil that has flowed down from the mountains.'

Living on the margin

More than 112 million people live in Bangladesh, in an area just a little bigger than England. It is one of the poorest countries in the world: 85 per cent of its people have too little to eat; 45 per cent have little or no employment. In the dry season there is too little water. In the monsoon, heavy rains, adding to the river waters already swollen by the melting snow of the Himalayas, bring frequent floods. Tropical storms are an ever-threatening and sometimes greater danger. Some 300,000 lives were lost in a single cyclone in 1970. In August and September 1988, monsoon floods killed an estimated 3,000 and made 25 million homeless. Most of the country's irrigation system was swept away, and at least 3 million tonnes of rice and other crops were destroyed. In December, a cyclone killed 5,000 more.

Researchers at the Woods Hole Oceanographic Institution estimate that a 1 metre rise in the sea level would cover 11.5 per cent of Bangladesh, an area containing 9 per cent of its population and 11 per cent of its agriculture, and generating 8 per cent of its gross domestic product. By the middle of next century, when a 1 metre sea rise is possible, the population of Bangladesh may have doubled. A 3 metre sea rise would drown 27.5 per cent of the country, with 27 per cent of the population and 20 per cent of the GDP. It would also put a larger additional area, including the capital, Dacca, at increased risk of flooding and saltwater intrusion of croplands and drinking water.

With climate change will come not only rising sea levels, but also an increase in the frequency and severity of storms. 'When one of those cyclones comes whirling up from the Bay of Bengal,' says Jules Pretty, of the IIED, 'a third to half of Bangladesh could be under the sea.' The Worldwatch Institute says:

> No one knows how many of the rice growing river deltas and floodplains in Asia or of the world's low lying cities will be inundated over the next century. The coastlines where protection might prove necessary in decades to come could easily total thousands of kilometres. The cost of trying to protect productive land and cities from the rising sea could be measured in trillions of dollars.

Feeding the five billion

Years of drought and famine in sub-Saharan Africa, where rainfall is erratic at the best of times and has been declining over the last fifteen years, have focused world attention on the vulnerability of tropical countries to variations in climate. These variations will become more frequent and less predictable as global warming takes hold, putting agriculture and the population at even greater risk.

Theoretically, and sometimes in practice, most of the developing countries within and beyond the tropics can feed themselves. The FAO has estimated that the 117 developing countries could produce enough food for one and a half times their projected population in the year 2000, even if they used no fertilisers or chemicals, planted traditional varieties of crops, and did nothing to protect the land from erosion and other forms of degradation. With some change in eating habits and improvements in agricultural efficiency the world could feed 10 billion people, the Brundtland Report concluded.

With the help of chemical fertilisers and pesticides, new seed varieties, and irrigation, food production increased by two and a half times between 1950 and 1986,

outstripping population growth. 'Yet 730 million people
still do not eat enough to lead fully productive lives,' says
Nafis Sadik, of UNEP. 'Two-thirds of them live in South
Asia and one-fifth in sub-Saharan Africa, mainly on land
which cannot support them by subsistence agriculture
alone.'

The growth of food production in these regions is
increasingly destroying the land resources on which
future output depends. The FAO estimates that without
conservation measures, soil degradation and erosion will
destroy 544 million hectares – 65 per cent of the cropland
which depends on rainfall in Asia, Africa and Latin
America – by the year 2100. In sub-Saharan Africa, about
65 million hectares of poor but productive land have
turned into desert in the last fifty years. In Sudan, the
desert has spread southward at the rate of 100km in
seventeen years. In Mali, the picture is even worse: a
350km advance in twenty years. According to Mostafa
Tolba, executive director of the UNEP, three billion hectares of the world's arable and grazing land – that's one
quarter of the world's land surface – is at risk from
desertification; 60 per cent is already either moderately or
severely degraded.

Irrigated farmland, which makes up some 15 per cent
of the total worldwide, but 30 per cent in India, 60 per
cent in China, 65 per cent in Pakistan, is already at risk
without the complications of rising temperatures,
changed rainfall patterns, and increased climatic unpre-
dicability. The FAO says that up to half of it is in danger
of becoming unusable, because of the twin evils of in-
adequately managed irrigation: the accumulation of salts
left behind as the water evaporates and waterlogging as a
result of insufficient drainage.

The vanishing forests

Most of the developing world depends heavily on organic matter – wood, and to a lesser extent, crop residues and dung – for its fuel. In sub-Saharan Africa, wood provides between 60 and 95 per cent of the fuel used for cooking, heating and even lighting, in cities as well as in rural areas. Over the last two decades, it has become apparent that in many places fuelwood resources are disappearing faster than they are being replaced. In the Rift Valley of Ethiopia, forest is being cleared for charcoal at the rate of 60,000 hectares a year. It has been estimated that India's forests are being cut at such a rate that they may disappear altogether by the end of the century. The FAO estimates that about 1.3 billion people live in areas where fuelwood is in short supply. If cutting continues at the present rate, the number may rise to 2.4 billion by the end of the century. Major aid programmes have been devoted to solving this problem (with variable degrees of success) by tree planting and the development of more efficient cooking stoves.

Recent studies in Africa by the IIED suggest that unreliable figures and crude projections of supply and demand have exaggerated the extent of the problem. The loss of forest owes more to clearance for agriculture than to fuelwood cutting. The report's authors, Gerald Leach and Robin Mearns, argue that the problem is trivial in terms of global energy consumption. If the woodburners were able to switch to more efficient fuels like oil and gas it would add the equivalent of only 9 per cent to world oil consumption. This is hardly a course that commends itself to anyone concerned about reducing carbon dioxide emissions from fossil fuel. Burning vegetation also produces carbon dioxide, but there would be no net addition to the atmosphere if the fuel were all taken from sustainably harvested wood or other crops, which can also be converted to more efficient liquid and gaseous fuels.

Natural forests cover some 9 million km sq of the

tropics. No one knows for certain how fast the clearance is proceeding: 110,000 km sq a year is the most commonly cited figure. But in 1987, some 82,000 km sq were lost in Brazil alone, and perhaps 130,000 km sq in the whole of the Amazon Basin, according to Norman Myers, a leading authority on the subject. The pace is accelerating, to the point where all Amazonia's forests could be lost within twenty-eight years. West Africa is losing 36,000 km sq a year, according to the World Bank. The pace of clearance appears to be speeding up in South-East Asia as well. The main cause is planned and unplanned clearance for agriculture, whether for subsistence or cash crops and – particularly in South America – for cattle ranching and land speculation. Commercial logging is an important contributory factor in Africa and Asia, its direct impact magnified by the opportunities it creates for farmers and settlers to invade otherwise relatively impenetrable areas.

Whatever the figures, they add up to a multiple environmental catastrophe. Tropical forests contain perhaps 80 per cent of all plant and animal species, many of them so far unidentified by modern science. It has been estimated that 10,000 species are being lost a year. The already established importance of many known plant species as sources of medicines and food crops makes the continued destruction of the forests a loss of incalculable value. Along with the forest species we are losing the subtle biological knowledge of the forest people. 'The Indians of the Amazonian Forest,' says Marie Paule Nougaret, reporting on a research project among the Kayapos whose homeland is threatened by development, 'select seeds, domesticate insects and practice a complex medical science that we are only just beginning to recognise.' The Kayapos distinguish 250 types of dysentry, and have a particular cure for each type. Each family is responsible for the culture and improvement of a specific plant, and they know how to select dozens of species that grow better together than separately.

Forest destruction alters the local climate by reducing the amount of water circulating between the trees and the clouds, and adds to the atmospheric burden of carbon dioxide, with Brazil alone contributing perhaps 10 per cent of the total man-made emissions each year. The land permanently cleared for agriculture – as opposed to the tribal method of shifting or temporary agricultural clearance – is rapidly dried out, exhausted or washed away. Replanting, where it is attempted, is mostly with single crops of commercial species such as eucalyptus: useless from the point of view of protecting biogenetic diversity and of little or no value to local people as a source of fuel or food. Concern in the developed world about forest loss is resented in the forested nations who argue that they need the land and the exports of timber, minerals and cash crops to support their growing populations. According to Jack Wentoby, formerly Senior Forestry Director at the FAO however, 'exploitation (of timber), with a few honourable exceptions, has been reckless, wasteful, even devastating . . . the operations have had no profound or durable impact on the economic and social life of the countries where they have taken place.' Unless ways are devised of making forest conservation a paying proposition for developing countries or changing economic policies to reflect the full costs of the forests, it is surely a lost cause.

Taking the strain

Great efforts are being made to counter environmental degradation in the developing countries, but progress is too slow to keep pace. The additional stresses of population growth and climatic change complicate the task while adding to its urgency. The impact of climatic hazards in developing nations is staggering, say Robert

The burden of debt

Poverty in many developing countries is a major cause of environmental degradation. Unless the causes of poverty and underdevelopment are rooted out, the over-exploitation and destruction of natural systems will continue. Nor will developing countries be able to join in a common effort to address the problem of global warming.

The Brundtland Report (see pp. 94–95) set out a programme for sustainable development which recognises that economic development and environmental protection are inextricably linked and must proceed hand in hand. The concept was endorsed by the United Nations and the 1988 Economic Summit of nine major western countries. Global warming, which had scarcely begun to emerge as a political issue when the Brundtland Report was published, makes that programme both more urgent and more difficult.

Over the past fifteen years or so, many developing countries have become poorer, and forced to draw even deeper on their environmental resources. The oil crisis of 1973, which made them pay much more for essential energy supplies, also obliged them to invest more of their own and borrowed capital in the exploitation of forests, mineral resources and the production of cash crops to earn foreign currency. Their people were forced into increasingly unsustainable farming and fuelwood cutting.

Developed nations, fearing a worsening international recession, encouraged higher borrowing. National indebtedness to commercial banks and international financial agencies spiralled: a three-fold increase between 1970 and 1976 alone. Although the pace of increase slowed in the 1980s, the total external debt of 109 developing countries almost doubled between 1980 and 1986 to over $1 trillion.

An even more damaging feature emerged as the price of commodities such as copper, iron ore, timber, rubber, cotton and sugar fell. The growth in developing world, foreign exchange earnings lagged behind the increasing burden of debt interest. Bankers became unwilling to finance repayments and interest by fresh loans, particularly to the more developed Latin American countries. In consequence, a net $140 billion was transferred from the poor to the rich nations from 1983 to 1988. In Brazil, one of the most heavily indebted countries, domestic inflation was 17,903 per cent in the forty-five months to December 1988. 'Growing poverty and deteriorating environmental conditions are clearly visible in every major Latin American country,' Brundtland reported.

The countries of sub-Saharan African continue to receive more in loans and aid than they return in interest and repayments, but the net benefit plummeted from $10 billion to $1 billion between 1982 and 1985. Per capita incomes actually fell by 16 per cent in five years. Austerity measures enforced by the International Monetary Fund cut into domestic spending programmes. According to the Brundtland Report:

> The critical situations in sub-Saharan Africa and the debt-strapped countries of Latin America demonstrate, in an extreme way, the damaging impacts that unreformed international economic arrangements are having on both development and the environment. If large parts of the developing world are to avert economic, social and environmental catastrophes, it is essential that global economic growth be revitalised. In practical terms, this means more rapid economic growth in both industrial and developing countries, lower interest rates, greater technology transfers, and significantly larger capital flows.

As efforts were made by the world financial community to stabilise the situation, banks began to sell back or exchange debts at discounts of around 50 per

cent and higher (98 per cent in the case of Sudan).
Western governments and international organisations
are now reviewing their strategies for dealing with the
problem. Interest rates have been reduced, and debts
partially cancelled in some cases. In others, debts have
been exchanged for shares in debtor country industries. The Soviet Union announced in November 1988
that it was prepared to write off some debts, and to
suspend debt repayment in other cases for up to a
hundred years.

In July 1987, the US environmental organisation
Conservation International signed an agreement with
the Bolivian government to buy $650,000 of Bolivia's
external debt at an 85 per cent discount, in return for
the protection of three areas of rainforest covering 1.5
million hectares. Several other 'debt-for-nature' swaps
have been agreed and the US Treasury has urged the
World Bank to support them with technical assistance,
financial advice and loans.

Chen and Martin Parry in a 1987 report for the UNEP and
IIASA:

In Africa, South America, and Asia, many countries
are plagued by persistent and recurring drought,
devastating floods and storms, damaging fires and
frosts, climatically-induced outbreaks of grasshop-
pers, locusts and other pests, highly variable fisheries,
and extensive desertification. At stake are the lives and
livelihood of millions of subsistence farmers, the
health and employment of tens of millions of urban
dwellers, the sustainability of diverse ecosystems, and
the solvency and integrity of national governments.

The Worldwatch Institute underlined the message in its 1988 *State of the World* report:

> To continue with a more or less business-as-usual attitude – to accept the loss of tree cover, erosion of soil, the expansion of deserts, the loss of plant and animal species, the depletion of the ozone layer, and the build-up of greenhouse gases – implies acceptance of economic decline and social disintegration. In a world where progress depends on a complex set of national and international economic ties, such disintegration would bring human suffering on a scale that has no precedent. The threat now posed by continuing environmental deterioration is no longer a hypothetical one. Dozens of countries will have lower living standards at the end of the 1980s than at the beginning. We can no longer assume that economic progress is automatic anywhere.

The developing world's perspective on these problems was made plain by Emil Salim, the Indonesian Minister for Population and Environment, and a member of the Brundtland Commission:

> Must we begin now to prepare for events perhaps fifty years ahead? We know that over this time-horizon the population of Indonesia is likely to expand from about 180 million to more than 350 million. The economy is highly dependent on fossil fuel for electricity generation and industrial growth, and upon tropical timber products for export. We are in need of rapid domestic industrial growth to provide employment, and food self-sufficiency is an ongoing concern. But our cities are rapidly expanding; most are located only a few metres above sea level . . . droughts have led to massive forest fires in Kalinmantan and shortages in reservoir water supply for major cities and for industrial irrigation use . . . millions of people are dependent on

the cultivation of shrimp and fish in brackish water fish ponds . . . We are not well prepared to cope with these potential problems, and we are not certain how much investment of our limited development budget should be committed to addressing this global problem which affects us all.

The rivers of discord

Access to water must be among the oldest sources of human conflict, and control of water one of the oldest sources of political and economic power. Today, eighty countries with 40 per cent of the world's population suffer serious water shortages, the Brundtland Commission reported. With water use expected to double by the end of the century, there will be growing competition for water for irrigation, industry, and domestic use. River water disputes have already occurred over the Rio Grande, in North America, the Rio de la Plata and Parana in South America, the Mekong and Ganges in Asia, and the Jordan, Litani, Orontes, and Euphrates in the Middle East.

Nowhere is the potential for conflict greater than on the Nile, whose basin is shared by nine nations: Egypt, Sudan, Ethiopia – where the Blue Nile arises – and, around Lake Victoria and the other sources of the White Nile, Zaire, Rwanda, Burundi, Kenya, Tanzania and Uganda. Egypt is completely dependent on the irrigated Nile Valley and Delta, an area of only 30,000 km sq in a land of more than 1 million km sq. Since 1970 it has regulated the flow by the Aswan Dam and Lake Nasser. The flow of the Blue Nile, once the greater source of water below the confluence at Khartoum, has been reduced by years of drought. At first this was offset by an increased flow in the White Nile. Now the waters of Lake Victoria have fallen, and Lake Nasser is at its lowest level for thousands of years.

Egypt and Sudan have had a joint agreement on Nile water since 1929, and a Permanent Joint Technical Commission to oversee sharing arrangements. They are jointly financing and constructing the 350km Jonglei Canal through the Sudd swamp of Southern Sudan where much of the river's water is lost. The project, begun in 1978, has been halted by political unrest in Sudan. The canal threatens many wildlife species and the livelihood of up to 400,000 Dinka, Nuer and Shilluk tribespeople, and may spread water-borne diseases. Both Sudan and Egypt are critically dependent for Nile water on Ethiopia. According to the Worldwatch Institute:

Although no major Ethiopian diversions of Blue Nile waters are foreseen, even a small diversion could cause serious problems. Sudanese needs for water may soon exceed its supply – after Egypt receives its share – and Egypt too is hard-pressed for water to support the rising agricultural needs of its burgeoning population.

UN attempts to bring the other nations into a sharing agreement have so far failed, and with upstream states likely to lay claim to Nile waters for hydropower and irrigation 'the greatest challenges would seem to lie ahead. There are few agreements either on twelve other African river basins each shared by four or more countries.'

In the Middle East, control of the Jordan was one of the purposes of Israel's occupation of the West Bank and the Golan Heights in 1967. Turkey controls the waters of the Tigris and Euphrates, which flow from the Turkish mountains through Syria and Iraq, forming the Shatt el Arab confluence which divides Iraq and Iran. Turkey has begun an irrigation and hydro-electricity scheme which involves damming the rivers, and is discussing water quotas with Iraq and Syria. It has also proposed a 'peace pipeline' to carry drinking water from two other rivers, the Seyhan and Ceyhan, which flow to the Mediterranean, through the Middle East to Saudia Arabia and the Gulf States.

Bangladesh, which consists largely of the delta of rivers flowing from the Himalayas, suffers from both too much and too little water. Control of the Ganges and Brahmaputra rivers, which converge in Bangladesh, is a long-standing political issue between India and Bangladesh. A barrage across the Ganges at Farakka, just above the border, enables India to divert the waters of the main river down the Hooghly Channel, which flows through India to Calcutta. Both countries want more of the low season flow, India for its crops, Bangladesh to repel saltwater from the Bay of Bengal and to stop the river channels becoming too silted up, and therefore more liable to flooding. A joint commission regulates the flow of the Ganges but has failed to resolve the political differences. Indian proposals for a canal from the Brahmaputra, through Bangladesh, to divert waters to the Indian section of the Ganges, have been blocked by Bangladesh (see also 'Living on the Margin', pp. 112–113).

The floods of 1988, which killed 3,000 and put 75 per cent of the country under water, were widely blamed by Bangladeshis on India, although there is little evidence to substantiate the claim that India could operate the barrage to prevent this almost annual event. In spite of the ill-feeling, the Indian Prime Minister Rajiv Gandhi and Bangladesh President Hossain Mohammed Ershad agreed in September 1988 to set up a joint task force to explore ways of preventing the floods.

Crossing the threshold

All they thought about this earlier was that there would be a slow, gradual change in ozone due to whatever it was we put

The plight of one in five

Slightly more than one fifth of the world's population lives in China, a land accustomed to natural disasters. Just over half the country is classified as arid or semi-arid. The desert area has increased by about 17 per cent since the 1950s to 176,000km sq, and is growing by 1,560km sq a year.

Droughts and water shortages are endemic, and getting worse. Water resources per capita are only about one-quarter of the world average: in North China the 1988 figure was put at 4 per cent of the world average. Exploitable water resources in the region were said to be 10 per cent short of demand. The Water Resources Minister Yang Zhenhuai, announcing new investment plans in November 1988, including flood-control projects and irrigation schemes for 50 million hectares of farmland, said that water facilities had been neglected for years.

A massive afforestation programme, aimed partly at halting the spread of deserts, is being undermined by pest epidemics. In 1987 about 5.5 million hectares were planted but 6.7 million hectares were infected. There were also serious fire losses. About 8 per cent of the country's 33 million hectares of pine forests are infested with pine moth, which is difficult to eradicate.

In 1988, droughts which reduced and in some cases destroyed wheat and rice crops on more than 11 million hectares – about 11 per cent of the total croplands – were accompanied by summer floods in many areas. These claimed 1,406 lives and drowned 7 million hectares of farmland. Around 60 million people were said to be affected and losses were put at billions of yuan (£1 = approx 6.6 yuan). By November 1988, an estimated 80 million rural people were threatened with food shortages, of which 20 million were said to be facing possible starvation.

More than 70 per cent of China's electricity is produced from coal, of which it has almost half the world's proven reserves. It is responsible for the third largest emissions of carbon dioxide, after the US and USSR.

China has one of the world's most rigorous birth-control programmes, but recently fertility rates have begun to rise again, significantly affecting global population forecasts.

into the amosphere. It turns out that it is not like that at all. It can go horribly non-linear on your at some stage and, if you do not understand it – and, my goodness, I have to say that we do not understand it in that detail – runaway effects can occur at almost any time

Joe Farman, discoverer of the Antarctic ozone hole, 1988

Joe Farman's evidence clearly impressed the members of the Commons Environment Committee, as they drew a parallel of this 'threshold effect' with other environmental problems such as acid rain and water pollution. 'We found from our . . . earlier inquiries that there is a degree of tolerance in so far as the physical environment is concerned, but once a threshold is reached and passed then disaster strikes very rapidly indeed.'

Every time that air, water or land pollution passes the medium's capacity to absorb, disperse and nullify the ill-effects, a preliminary threshold is passed. The environment is set on a course which, if unchecked, may bring it sooner or later to a second, more catastrophic threshold. That may be happening now in the declining forests of Europe and North America through some combination of acid and photochemical pollution, disease and climatic stress. It is happening where over-exploitation reduces land to desert or leaves the soil waterlogged or salinated.

It may be happening, environmentally and socially, in the world's expanding mega-cities.

The addition of man-made greenhouse gases to the atmosphere beyond the capacity of natural systems to absorb and recycle them represents another threshold crossed. The planet is already committed to some degree of warming. We cannot be sure that it will not, if un-checked, bring the atmosphere nearer to another more catastrophic change. Nor can we tell how and when the resulting climatic change may take some already over-stressed natural or human system across the threshold. For some species and habitats, lacking time and space to adapt, the future is bleak indeed. For the rest of us, can we afford to wait to find out?

World leaders have begun to lay aside at least some of their differences to speak in the language of common determination. The message has already begun to strike home, if not yet with sufficient urgency. We must begin to take action to reduce global warming. And we must learn that global warming is in reality a global warning to heed the thresholds of ecological tolerance. The massive experiment is going wrong. As Eduard Shevardnadze argued in his speech to the United Nations: 'the second front' is all around us.

5

MAKING THE CHOICES

> Don't shoot till you see the whites of their eyes.
> British Army Saying

Where do we stand now as we face the second front of environmental destruction? It may be the greatest battle humanity has ever taken on. It is the first in which we may hope to be united in the face of a common enemy. The threat confronting us is the degradation of our environment to the point where it is becoming decreasingly capable of supporting us. This we know to be the consequence of our own past ignorance and carelessness. The best hope of success, the ground of all our optimism, is that we have at last recognised the enemy.

The ideal strategy for dealing with any threat is to choose the exact moment when action has the best chance of success without waste of effort. We must, of course, have good knowledge of the problem, a plan of battle, a range of tactics, adequate resources, and a will to win. If we define the problem purely in terms of global warming, the strategic moment is far from clear. Substantial doubt remains, not about the existence of the threat but about the speed of its advance and the nature of its impact on our lives and the environment. Are we, then, at a wait-and-see stage in which the wisest course is to gather intelligence, prepare our battle plans, put ourselves in training, but save our ammunition until we can 'see the whites of their eyes'?

David Everest, former Director of Environmental Science Policy in the UK Department of the Environment, set out the issues and options for the public policy-makers in a paper published in October 1988 by the Joint Energy Programme of the Policy Studies Institute and the Royal Institute of International Affairs. He placed his discussion in the context of 'political considerations . . . the ultimate arbiter of policy.' To ignore political considerations, he wrote, 'would only result in the formulation of unsustainable policies in a field where long-term stability is of the greatest importance.'

Everest's first option was 'do nothing and await the results of further research'. This, he observed, best summarised the current approach of Britain and most other countries:

> There is no present evidence which would point to the interests of the UK being damaged by a limited greenhouse warming. Considerable costs would be involved in taking needless precautionary action, say by increasing the size of sea defences, whilst measures to control CO_2 emissions directly would place very serious constraints upon the free working of the energy market, to which the present British government is committed.

It was important politically not to jump the gun with policies which involved high expenditure or foreclosed other options, Everest argued. The wait-and-see option, however, was most plausible if one considered that any greenhouse warming was likely to be at the low end of the predicted range and that feedbacks not represented in the models were as likely to reduce as increase any warming. Doubt over whether past increases in greenhouse gases had led to any warming lent credence to this approach. A slow onset of limited warming would appear to give time for adaptations.

There were risks, however. Energy consumption, and

therefore carbon dioxide emissions, might rise faster than present trends indicated; the warming might be at the higher end of the predicted scale; feedback mechanisms might amplify the warming. Basing coastal protection policies on current rates of sea-level rise rather than those predicted might underestimate the risk of flooding. Nevertheless, Everest concluded, the approach was likely to endure as long as there was no significant change in the scientific consensus or no other political pressure related to the greenhouse effect.

Everest suggested that one such political pressure could arise 'if the British government saw political advantage in being up with the leaders on this issue, rather than being perceived internationally as being back with the laggards, as occurred in the case of the acid rain issue.' This thought seems to have occurred to Margaret Thatcher even as Everest was writing. Her Royal Society speech was delivered a few days before Everest's report was published. She also put in hand a departmental policy review. Whether that will result in a more active approach to the global warming problem remains to be seen.

A variation on the 'do nothing' approach is favoured by the veteran Soviet climatologist Mikhail Budyko, who believes that for the next few decades global warming may improve agricultural prospects. If this were so, he told the Climate Change and Development Congress in Hamburg in November 1988, then attempting to curb global warming would be 'not only useless but even dangerous'. International agreement to reduce carbon dioxide emissions could take decades, Budyko argued. It might be better to increase emissions to hasten the beneficial warming. By the time it became an urgent problem the shortage of economically recoverable fossil fuels might do the job of reducing emissions of its own accord.

Budyko's claim that fossil fuel shortages will eventually do the trick has some support, but his complacency is not shared by the scientific community at large. Nor is

A summary of uncertainties

Predictions about global warming depend, as we have seen, mainly on computer models which mimic the response of the atmosphere to an increase in the greenhouse gases. There is no doubt that the greenhouse effect works, but there are uncertainties about the amount of warming and sea-level rise which will accompany a doubling of greenhouse gases, and about the speed of change.

Uncertainty about the speed of warming mainly springs from the role of the oceans as reservoirs of solar heat. Not enough is known about the rate at which they absorb heat, move it about, and return it to the atmosphere. The lag involved in this temporary storage and delayed delivery could be anything from fifteen to fifty years. Not enough is known either about how much extra carbon dioxide the oceans are able to absorb and lock up, particularly when it is itself warming.

Uncertainty about the amount of warming is related in part to our inability to know – rather than predict – how much extra carbon dioxide and other greenhouse gases will be added to the atmosphere in future years. It also arises because of an incomplete understanding of feedback mechanisms affecting the concentration of greenhouse gases and the response of the atmosphere. Warming may result in more gases being emitted because of the melting of the tundra and the destruction of forests. This might be slightly offset by more vigorously growing plants absorbing more carbon dioxide. Reduced ice and snow will mean that less solar heat is reflected back into space; but a reduction in forests and an increase in deserts would lead to more heat being reflected rather than absorbed. Changes in the amount and type of cloud can act both ways: trapping more heat in the atmosphere but also reflecting more sunlight back into space.

One way of testing the reliability of the computer

models is to see how close their results come to the known facts when the programme is run over a past period. The Goddard Institute modelled the greenhouse effect likely to have occurred since 1958, when accurate measurements of atmospheric carbon dioxide began. Although there were 'major uncertainties' of the type described above in the model, the results were almost identical to the observed warming over the period to 1987 – about 0.4°C. They also followed the annual variation closely over most of the period, except in the mid-1970s when the observed temperature fell much more sharply than predicted. A warming of 0.4°C over thirty years is four times the amount the temperature normally fluctuates about the average over a thirty-year period. The probability of a *normal* deviation of this magnitude is about 1 per cent.

When run forward, the model predicted a temperature rise of 0.6°C by the year 2019 if drastic efforts were made to reduce emissions, and a rise of 1.5°C if emissions continued to rise at the rate experienced over the last twenty years. It predicted an increased chance of droughts and heat waves in the south-east and mid-west United States in the late 1980s and 1990s.

James Hansen and his Goddard team programmed the model to predict the likelihood of hot summers in two areas of the USA: Washington DC and Omaha. They divided the summers between 1950 and 1979 into the ten hottest, the ten coldest, and the ten 'normal' years. By comparison with these years, the model predicted that by the 1990s, both areas would have a more than 50 per cent chance of a hot summer.

Hansen translated the odds into gamblers terms. Take a die and paint two faces red and two blue for the hot and cold summers in the 1960s and 1970s. By the 1990s, the model has painted three or four of the faces red. 'It seems to us that this is sufficient "loading" of the die that it will be noticeable to the man in the street.' Hansen says. What is more, the model leaves one face blue in the 1990s and the first decade of the next century: there is a substantial chance, therefore, that hot summers will be mixed with cold ones.

'Although the greenhouse effect is usually measured by the change of mean temperature,' Hansen says, 'the frequency and severity of extreme temperature events is probably of greater importance to the biosphere.' Both plants and animals are affected by extreme temperatures, and the range of local temperatures often defines whether or not a place is habitable.

there any evidence that the USSR is guided by his views. Indeed, Vladimir Klimenko from the Moscow Power-Engineering Institute has described Budyko's Hamburg Congress paper as like 'swearing in church'. There is concern, however, that this kind of attitude by those who see potential national advantage in global warming will prolong international negotiations.

The 'wait-and-see' approach has obvious attractions, particularly to those who lack the present means to act ahead of strong evidence. We note such an instance in the Anglian Water Authority's attitude to sea defences (see chapter 3). It must be an attractive course for those developing countries which see themselves as the innocent victims of something visited on the world primarily by the industrialised nations. It may be the prudent course in some cases; but it carries at least three risks. The first is that the cost of taking action may be higher the longer it is delayed. The second is that some of the solutions now available may disappear if action is left too late. The third is that action taken in the face of urgent necessity may prove short-sighted and harmful to other environmental, social and economic interests.

Richard Warrick, of the University of East Anglia's Climatic Research Unit, cautions against taking such risks. 'We are forced,' he told a Royal Society of Arts Conference on the Future of the Countryside, 'to make decisions in the face of extreme uncertainty, and we have to start making them now. We are going to have to start using some best guesses rather than wait for certainties.'

Kenneth Hare, chairman of Canada's Climate Program Planning Board, put the point even more forcefully at the Toronto Conference in June 1988. The broad conclusions of the climate modellers, although still clouded with uncertainty, were at least as firm as the projections of economic forecasting models. 'If decision-makers are willing to listen to economists, they should be even readier to listen to the natural scientists as regards future outcomes.'

Writing in the science journal *Nature* a couple of months later, Hare admitted that he had been until recently 'an announced wait-and-see conservative'. He was tipped towards certainty, he went on, by the evidence of an unmistakable unward trend in global temperatures since 1860. This was not proof, he wrote, but it fitted the predictions.

> As a scientist, I will hence confine myself to saying that the best available explanation for the upward trend of surface temperature is the build-up of greenhouse gases . . . As an adviser to my government I have to be more explicit. I can and do tell them that they should base their environmental planning on the assumption that the greenhouse warming will continue and accelerate. There will always be conservatives who decline to go this far. At the age of sixty-nine I can no longer afford to be a conservative.

Hare was writing in response to an editorial from *Nature*'s John Maddox who was sceptical that global warming had been proven. A few months later, in an apparent change of heart, Maddox told an international ozone conference in London that 'despite the uncertainties that persist, it is by no means too soon to begin embarking now on the negotiations and agreements that will be necessary to regulate the "greenhouse" problem.'

Searching for clues

A logical extension of the wait-and-see policy, David Everest wrote, was to increase the amount of research and thus attempt to speed up scientific certainty. Such an approach had political benefits for the UK, he argued: it would demonstrate concern and could increase Britain's expertise and influence in the field. The government's record in this area is less respectable than it might be. It cut research funds in the early part of 1988, just as global warming was beginning to emerge as a public issue in the UK. The Natural Environment Research Council (NERC), which directs much of the UK's global warming research, had to write off two unique ocean research programmes. One involved the role of plankton. The other concerned a still unexplained phenomenon, an increase in the average height of waves in the Atlantic.

NERC's then chairman, Hugh Fish, warned that there were 'huge gaps' in scientific understanding of the role of the oceans, both in taking carbon dioxide from the atmosphere and as a major influence on climate. The Council was involved in many collaborative international research endeavours, directed at filling this gap, he wrote in its annual report. The extent to which it could proceed had been put 'in serious jeopardy' by the cuts in funding.

The UK House of Commons Environment Committee urged additional research funding in a report on atmospheric pollution in May 1988:

We are disturbed about the lack of scientific knowledge
on the greenhouse effect several decades after it was
first anticipated as a problem. We are particularly
concerned about the lack of information on the detailed climatic and ecological implications for the UK.
The anticipated lag time between the emission of
greenhouse gases and experience of climatic change
means that it is vital to understand the likely future
impact of gases currently being released . . . Unless we
take steps now we may have to live with the consequences thirty years down the road and . . . trying to
deal with them twenty years from now might be too
late.

Responding to the report in December 1988, the government's Environment Minister Lord Caithness said:

The first priority must be to improve scientific understanding by co-ordinated international research so that
long-term policies can be based on firm foundations.
The UK is playing a significant part in this research
through the Meteorological Office's global climate
modelling programme.

The Environment Department restored some of the
NERC's grant and announced additional research con-
tracts. It also took the chairmanship of an international
team set up to review scientific knowledge on climate
change.

One NERC agency with a strong involvement in climatic change research, the British Antarctic Survey
(BAS), is financially protected by the government, chiefly
because of the political importance of the Falkland
Islands and the potential economic importance of nearby
Antarctica. In addition to its work on the ozone hole, BAS
is involved in studies of the west Antarctic ice sheet and
currents in the Southern Ocean. Other NERC research

programmes related to climate change include monitoring of heat, water and momentum exchanges between the oceans and the atmosphere and the role of the oceans in absorbing carbon dioxide. It is also involved in a major European research programme on sea levels, aimed at producing computer models capable of predicting sea-level trends and changes in storm surges. Maps will be produced showing the settlements, farmland, industrial installations and natural ecosystems at risk.

No one doubts that more research is needed across a wide field, from the behaviour of the oceans to the impacts of climate change on the environment and society. We set out some of the priority areas in chapter 7.

Changing our ways

Everest's next category of response, adaptation to climate change, is not so much an option as an inescapable necessity. We are committed to some degree of global warming by the excess greenhouse gases already in the atmosphere; and the concentrations will continue to grow for some time, even if we act with all practical speed to curb them. Everest argued that adaptation 'must be an element, possibly the dominant element, in any policy mix relating to the greenhouse effect.' Rising sea levels required an increase in coastal works, modifications of port systems, and the provision of water supplies to replace those lost by salt penetration. Agriculture would need more irrigation, new or modified farming practices, and new crop strains.

'If the warming is relatively limited and slow to emerge, then adaptation to its impacts might well be left to the normal evolution of society, including the action of market forces,' he wrote. 'Even a significant greenhouse warming could well be met, at least partly, through adaptations: indeed it might be the best overall re-

The precautionary principle

Do we have to wait for positive proof of global warming before taking action to reduce the emissions of the greenhouse gases? Acting on suspicion is a well-established principle in pollution control. It is not, traditionally, the British way of doing things. The approach of successive British governments has been to demand sound scientific evidence, a clear chain of cause-and-effect, before introducing or strengthening pollution controls.

The government's argument has always been that it is not prepared to spend scarce resources on solving an unproven problem when there are other, proven problems urgently demanding attention. It has, however, begun to modify this attitude.

At a ministerial conference about the North Sea held in November 1987, the government stated that it accepted the German and Dutch argument for adopting a precautionary approach to pollution control. Under the North Sea agreement, action will be taken to limit the discharge of pesticides and other substances which are persistent, toxic and accumulate in the food chain, posing a threat to fish, birds and sea mammals, and possibly to people eating seafood. The agreement also phases out the incineration at sea of toxic chemicals.

These actions are being taken on evidence that points to, rather than proves, environmental damage to marine life. The government proposes to write the precautionary approach into all future pollution-control legislation.

The North Sea Conference was a step forward in international co-operation on pollution problems. But a great deal remains to be done even in such basic areas as agreeing common standards for monitoring the amount of pollutants in the sea. Even in such a small part of the globe as the North Sea, and among willing partners, it is taking a long time to mount effective co-ordinated action to tackle a commonly perceived environmental problem.

sponse.' But it was a potentially expensive strategy and not without risks, he added, because the measures taken might prove insufficient or excessive. An adaptative approach should go hand in hand with additional research and international discussion of the underlying science.

Some adaptation will occur without us really noticing. It is happening all the time: changes in diet, in lifestyle, working habits and recreational activities. Choosing deliberate adaptation policies, however, will be far from straightforward. There is an inherent risk – as indeed there is in any policy response – that new approaches intended to cope with the effects of climate change could exacerbate either climatic change itself or other forms of environmental degradation. We examined some of the issues involved in coastal protection in chapter 3. There are also problems with agriculture.

Changed climates may dictate increased use of fertilisers to maintain or increase food production; but nitrogenous fertilisers are a source of nitrous oxide and a cause of water pollution. Climatic change may dictate increased use of pesticides; but pesticides pollute water, damage wildlife, and contaminate food. This dilemma may be particularly acute in developing countries which are already struggling to win a living from increasingly degraded land. Some of their existing practices, however, such as forest clearance, are contributing to the burden of greenhouse gases. Consequently, improved land management is needed urgently, regardless of climatic change.

Damaging land use practices in developing countries do not arise from ignorance or stupidity, as Commonwealth Secretary-General Shridath Ramphal points out:

There is, in most poor countries, a sophisticated awareness of the kind of agricultural practices that are sustainable. In India, China, Indonesia and also in many parts of Africa, there are, in peasant farming communities, traditions of terracing, crop rotation,

natural fertilisers and animal husbandry that long pre-
date the arrival of European technology. But poor
countries often find themselves trapped in a downward spiral in which the combined pressures of poverty and rising population lead to sound practices
being abandoned.

With the help of aid agencies and research organisations,
many traditional practices are being revived.

In the drought regions of Africa, says Lloyd Timberlake, of the International Institute for Environment and
Development, awareness of climate change is a hin-
drance to much needed land use improvements:

It tempts leaders into apathy. Many of the district
officers, forestry and range management workers I
have talked to in the Sahel see themselves and their
people as victims of an angry god or a changing climate. Some Sahelian officials concerned with the
countryside tend to throw up their hands and say: 'The
climate is changing. So what can we do?'

Timberlake points out that droughts have been occurring
in the Sahel for centuries:

For drought-stricken Africa, the cry that the climate is
changing is ultimately a cop-out, an excuse for political
inaction. We do not know if the climate in dryland
Africa is really becoming drier, and we do not know
how to reverse this change quickly even if it is occurring. But it is certain that bad land management is
reducing the use that can be made of the rain that does
fall, causing the moisture to evaporate or run off
damaged soils rather than seep into the ground to be
used by crops and vegetation . . . Misuse of the land is
widespread, increases vulnerability to drought, and is
reversible. Rainfall patterns cannot at present be either

modified or predicted, but human behaviour can and must be changed.

Turning down the heat

The final option in Everest's analysis is action to moderate the greenhouse effect. Like adaptation, this is not a choice we can ignore. The issue is not whether we should act, but when, and how.

The world's sudden awakening to the greenhouse threat in 1988 produced some wonderful and occasionally bizarre ideas, from destroying all the termite nests (termites produce methane) to adding sulphur dioxide to the oceans to stimulate the growth of algae (which remove atmospheric carbon dioxide by photosynthesis). Other bright ideas included floating billions of polystyrene balls in the oceans and spreading dust in the stratosphere, both intended to bounce sunlight back into space. Ozone depletion, it was suggested, could be overcome by injecting more ozone into the atmosphere from high-flying aircraft.

Another set of proposals seeks to offset sea-level rise by storing more water on land. These are, in effect, new justifications for old schemes to divert major rivers. The Soviet Union has long-standing ideas about refilling the Caspian and Aral seas from Siberian rivers like the Lena and Ob. In North America there is a proposal to transfer water from north-west Canada to the Colorado river by way of a huge reservoir in the Rockies. The 'Rocky Mountain Trench' might save two years' worth of sea-level rise.

Such 'dangerous technological options' should be considered the absolute last resort, says James Lovelock. 'The cure might be worse than the disease,' says Stephen Schneider. 'The prospect for international tensions resulting from such deliberate environmental modifications is so staggering, and our legal instruments to deal

with it so immature, that it is hard to imagine acceptance
of any substantial mitigation strategies in the foreseeable
future.'

On a more practical level, action has already begun
to reduce the emission of CFCs. The plan agreed at
Montreal in 1987 (see 'The Montreal Protocol', pp. 143–
145) to protect stratospheric ozone has the added advantage of tackling the most potent of all greenhouse gases.
It needs urgent amendment to be effective in either role.
If current diplomatic efforts to tighten its restrictions and
bring non-signatory nations within its scope succeed,
however, it will represent an important break-through.
Uncontrolled, CFC emissions would represent up to a
quarter of the global warming commitment inflicted on
the planet over the next fifty years. The achievement of a
substantially complete world-wide agreement to halt
emissions would have a greater significance as well. It
would present a model for action to curb the other
greenhouse gases.

No strategy for reducing the greenhouse effect can
work without action to cut carbon dioxide emissions. It is
the most abundant of the greenhouse gases and, after
CFCs, the most amenable to action. There are, in theory,
two ways in which it can be tackled. The first, however,
does not look a like a viable option. It is to treat carbon
dioxide like other troublesome fossil fuel pollutants and
remove it from the furnace chimney and the vehicle
exhaust pipe. However, carbon dioxide is not, like sulphur dioxide and nitrous oxides, a minor constituent of
the exhaust stream: it is the main product, besides heat,
of burning any carbon based fuel. Collecting it from car
exhausts, domestic buildings and other small sources
would be quite impracticable.

For power-stations and other large burners, the suggested method is to convert the gas, with solvents or heat
exchangers, to liquid carbon dioxide. It would then have
to be safely stored away somewhere. The disposal options include deep geological voids such as exhausted

The Montreal Protocol

The first practical step towards limiting CFCs was the United Nations Convention on the Protection of the Ozone Layer, drawn up in Vienna in 1985, which provided for international co-operation to tackle the problem. The Montreal Protocol, which came into force on 1 January 1989, is the actual agreement to reduce consumption of CFCs. It commits the signatory states to review the terms of the agreement every four years, starting in 1990.

Under the initial terms, each signatory nation is required to freeze and then reduce its production and consumption of five CFCs compounds and three halon compounds in accordance with a timetable. The base year for calculating the amounts is 1986.

Bulk consumption of CFCs (11, 12, 113, 114 and 115) is frozen at 1986 levels from 1989, reduced to 80 per cent of the 1986 level by 1994 and to 50 per cent by 1999. Production is subject to less severe cuts, largely to meet the needs of developing countries. Production in 1989 must be no more than 110 per cent of the 1986 figure, and must then be reduced to 90 per cent by 1994 and 65 per cent by 1999. Consumption of the three halons (1211, 1301 and 2402) is frozen at the 1986 levels, and production limited to 110 per cent of that level in 1992.

Since the compounds vary in their ability to destroy ozone – CFC115 is only 60 per cent as effective as CFCs 11 and 12, and 75 per cent of 114, for instance – the agreement allows production and consumption to be switched between the compounds providing that the total destructive potential is not changed.

Because most of the compounds are long-lived, the effects of existing concentrations on ozone and global warming will continue. Some of the uncontrolled compounds, which may be used in substitution, are also powerful greenhouse gases (see appendix 5).

The US EPA calculate that if every country adopted

the Montreal Protocol, the amount of chlorine in the atmosphere would still increase from the present level of about 2.7 parts per billion by volume (ppbv) to between 6 and 8ppbv by the year 2075. Immediate elimination of all fully-halogenated compounds and a freeze on methyl chloroform is needed to stabilise the stratospheric concentrations at present levels over the next hundred years.

Without the Montreal Protocol, CFCs would add about 50 per cent to the global warming effect of carbon dioxide in the next few decades. With the present terms of the Protocol in force, this could be reduced to 7 to 16 per cent. Permitted substitutes could, however, push the figure up again. The compound HCF134a, which is being developed as a permitted replacement for CFCs in fridges and air conditioners, would be 20 per cent as effective as the CFCs in causing global warming if emissions grew by the 3 per cent a year originally expected for CFCs.

According to Keith Shine, a member of the UK Stratospheric Ozone Review Group, if all the controlled CFCs were replaced by HCF134a or something like it, and emissions were not controlled, the result could be a committed global warming of nearly 1°C within a century. These are admittedly 'back-of-the-envelope' calculations, but on this basis the CFC substitutes would be in the same greenhouse-effect league as methane and nitrous oxide.

Many governments have already heeded the warnings of the scientists and are pressing for faster and more effective Montreal controls. The British government, which initially tried to block moves for stringent controls, says it expects the 50 per cent reduction target will be met in the UK by the end of 1989 and that cuts of at least 85 per cent should be scheduled as soon as possible. The Swedish government has scheduled a 50 per cent cut by 1991 and a virtual phase-out by 1995. The Dutch and Canadian governments are heading for a complete phase-out by the year 2000 and 1999 respectively. West Germany looks like committing itself to a 95 per cent minimum reduction by 1996. As this book went to press, the EEC agreed in principle to 'at least' 85 per cent cuts and a virtual phase-out by the turn of the century.

coal and salt mines and oil wells – all of which might prove to be leaky – and the ocean depths. For ocean disposal the liquid would have to be piped to a depth of about 3,000 metres. At that pressure carbon dioxide is denser than sea water and it would, in theory, stay put. The possibility of several decades' accumulated carbon dioxide breaking to the surface cannot, however, be discounted. The whole operation would also be enormously expensive. Half the electricity generated in a power-station would be used to dispose of a single waste product.

The more feasible alternative is to reduce the amount of carbon dioxide emitted. One method is to switch from coal to oil and natural gas, both of which produce less carbon dioxide for a comparable amount of heat. Natural gas, the cleanest of the three fuels, is the most promising substitute. Everest described it as an attractive interim measure which would:

reduce the rate of emissions of the most important greenhouse gas on a precautionary basis, whilst allowing time for either improved scientific understanding of the extent of future greenhouse warming to emerge and/or the development of more comprehensive control policies.

As a short-term measure it faces the obstacle of a large, well organised and economically important coal industry. Coal is a major employer in some countries and has large amounts of capital sunk, so to speak, in mines. It is also, for many industrialised and developing countries (including China which plans to almost double its consumption in fifteen years) a domestically available source of energy. These are assets which cannot be switched on and off like a tap. Nevertheless, it represents for some countries at least a real opportunity to make a start on the carbon dioxide problem. In the UK, with its own offshore sources, the contribution of natural gas has increased from 5 to 25 per cent of the primary energy consumption since 1970.

The second option is to reduce in relative or absolute terms the amount of energy used. Can it be done? To ask for a cut in energy consumption when the world population is growing so fast, and so many are short of fuel, food, and shelter, sounds nonsensical. Global energy consumption in 1980 was around 10 Terawatts. (A Terawatt (TW) is a billion kilowatts, and the equivalent of burning about 1 billion tonnes of coal.) People in the industrialised world use much more energy than those in the developing countries – on average, more than six times as much. Continuing to provide energy at the present rate would increase global consumption to 14TW by 2025 when the population will be over 8 billion. To give everyone as much energy as the industrialised nations use would raise the figure to 55TW.

The World Commission on Environment and Development (see 'The Brundtland Report', pp. 94–95) examined a number of forecasts of how energy demand might grow over the next few decades. One, by the International Institute for Applied Systems Analysis (IIASA), put the figure at 35TW by 2030. This would require about 1.6 times as much oil, 3.4 times as much natural gas, and nearly five times as much coal as in 1980, and a thirty-fold increase in nuclear energy – opening a new nuclear power-station every two to four days. It would involve the serious probability of climate change, as well as

growing air pollution, acidification and the risk of
nuclear accidents. Similar projections have been produced by the World Energy Conference (WEC).

Another forecast, by a team from the US, Brazil,
Sweden and India, assumed a demand of 11.2TW in
2020. Even this projection, which requires the industrially advanced countries to halve their energy consumption while allowing a 30 per cent increase in
developing countries, involves some commitment to additional global warming. It also requires 'an energy efficiency revolution' which the Commission thought most
governments were unlikely to fully realise within forty
years. It would, however, permit an annual global increase in per capita gross domestic product of about 3 per
cent – the figure regarded by the Commission as essential
to provide for people's needs.

Unsurprisingly, both projections have their critics. The
IIASA study was a massive effort, involving 140 scientists from twenty countries in a seven year research
programme which cost $6.5 million. Amory Lovins, Di-
rector of the Rocky Mountain Institute, says 'virtually
identical results . . . can be reproduced in a few minutes
on a $4.95 hand calculator.' Two former IIASA re-
searchers, Bill Keepin and Brian Wynne, observed
that many of the inputs to the Institute's scenario were
'based on opinion rather than objective robust analysis.'
Thomas Johansson, from the University of Lund in
Sweden, says the IIASA and WEC scenarios are 'fatally
flawed'. They assume a close coupling between a nation's energy demand and its level of economic activity.
They 'simply extrapolate general patterns from the past
into the future without any attempt to understand the
lessons of the past or analyse the potentials for energy
efficiency.'

Sceptics questioned whether the WRI scenario, based
on the energy-saving example of the most advanced
economies was applicable in other, less advanced, industrialised countries, let alone in the developing nations.

But the team, led by Jose Goldemberg, President of the University of Sao Paulo, point to the hidden reality of energy consumption in the developing countries. Most of the commercial energy, they argue, is used by the elite class of industrialists, businessmen, officials, politicians and professionals who make up no more than 10 to 15 per cent of the population. These elites they say 'have acquired the energy-using habits of consumers in rich countries, and often they waste even more energy'.

Low-energy scenarios of the type produced by the WRI focus on the 'end uses' – the various tasks such as cooking, lighting, heating, transport, and industrial process which need energy – and look for technical means of improving efficiency. One study in the UK, by the Earth Resources Research group, estimated that the country could reduce consumption by 30 to 60 per cent by reducing heat losses in buildings, using more efficient electrical appliances and motor vehicles, improving public transport, and doubling the efficiency of power-stations. Studies in the US, Canada and West Germany suggest potential savings of up to 50 per cent.

Energy efficiency requires fundamental political and institutional shifts, the Brundtland Report admits. But it adds: 'The Commission believes that there is no other realistic option open to the world for the twenty-first century. The ideas behind these lower scenarios are not fanciful. Energy efficiency has already shown cost-effective results.'

The Commission's optimism comes from the lessons learned in industrialised countries after the rapid rise in oil prices in the early 1970s. In the five years before the 1973 oil embargo, the energy consumption of industrialised nations rose by 26 per cent. Since 1973, however, their energy consumption has been falling even though their economies have continued to grow. The 'close coupling of energy and GDP' assumed by IIASA has been broken. In Japan, already one of the most energy-efficient countries, the reduction in energy intensity (the amount of energy to provide a unit of economic output) was 31 per cent. In the US it was 23 per cent, in the UK 20 per cent, in West Germany 18 per cent.

Quite simply, faced with higher fuel prices, countless large energy users discovered that their industrial processes and their heating and lighting systems were grossly inefficient. 'Process and equipment changes make the average Japanese paper plant or steel mill 30 to 50 per cent more efficient than it was a decade ago,' the Worldwatch Institute says in its 1988 *State of the World* report. A new American office block has about the same lighting level and temperatures as older ones, but uses less than half as much electricity. The report adds:

> No country has even begun to tap the full potential for further improvements; a range of technologies now coming on the market are more efficient than experts thought feasible just a few years ago. Efficiency gains of at least 50 per cent are available in every sector of the economy. Yet from wood stoves in African villages to office buildings in California, the limiting factor is not technical but rather institutional.

(For further discussion of energy efficiency, see chapter 7.)

Energy efficiency has been called a 'quiet revolution', because no one at the time realised that it was happening. It was not a planned or legally enforced change but a spontaneous response to market forces: an example, if you like, of unconscious adaptation to changing circumstances. This is not to suggest we can count on the process continuing without exhortation or a legislative nudge in the right direction. Energy saving has become too important to be left to unregulated market forces. It is, the Brundtland Report says, the only practical course for meeting the world's growing energy needs.

A reduction in energy consumption, whether relative or absolute, offers more than a reduction in carbon dioxide emissions. Burning less fossil fuel means producing smaller volumes of sulphur dioxide, nitrogen oxides, carbon monoxide and hydrocarbons, all damaging to the environment and to health, and a smaller cost to remove them from chimney stacks and exhaust pipes.

The Brundtland Report has been endorsed by world leaders. A global strategy for curtailing the greenhouse effect and meeting growing human needs is, therefore, in the making. This strategy, and its implications for the international community, governments, corporations and individuals, is examined in chapter 7.

Clean energy

Using 'cleaner' fossil fuels and making better use of the energy that they provide will take us a long way towards curbing carbon dioxide emissions. The only viable long-term option, the Villach–Bellagio workshops concluded, was to replace fossil fuels with alternative energy sources. Those currently available are nuclear power (which is discussed in chapter 6) and the so-called 'renewables' – hydro-electricity; wind, wave and tidal energy; solar energy; biomass; and geothermal energy. With the exception of hydropower, which provides some 25 per cent of the world's electricity, and biomass, which provides 6 per cent of total energy supplies, the renewables are at present only minor contributors to the overall energy output. How practicable and desirable are they as the main element of a strategy to cut greenhouse gases? Since they depend on the forces, not to say the whims, of nature, they are all subject to some limitations of useful availability. There are also technical problems and environmental impacts to consider.

Hydro-electricity, particularly on a large scale, often involves flooding large areas; drowning potentially productive land, natural habitats, homes, and settlements. In the developing countries in particular these measures have become associated with the enforced migration of tribal people. Dams interfere with river flows, giving rise to potentially serious downstream problems. Where rivers cross national, or even regional, boundaries, this can engender political conflict (see chapter 4). In the eyes of some conservationists, large-scale hydro-electricity schemes rank close to forest destruction as an evil visited on developing countries by the developed world. On a smaller scale, however, hydro-electricity can make an important and cost-effective contribution without giving rise to these problems.

Wind, unlike water, cannot be stored to provide a steady source of energy. It has to be caught where it is found, and large numbers of turbines are needed to yield a commercially viable source of electricity. Extensive 'wind parks' may be regarded as visually intrusive and unwelcome in areas of fine landscape. Public response to wind parks in Denmark and the Altamont Pass in California has generally been positive however. Contrary to the impression given by some, the land can continue to be cultivated or grazed. There is increasing interest in locating wind parks in shallow offshore waters. Wind power is now making a growing contribution in Denmark and the US. In Britain, the Central Electricity Generating Board believes that electricity can be generated as cheaply by wind as by its latest proposed nuclear power-station at Hinkley Point in Somerset.

Tidal barrages on estuaries, currently under investigation in the UK, are liable to drown ecologically important mud-flats and marshes, and replace salt or brackish water habitats with freshwater ones. Their contribution to electricity needs would be around 20 per cent if six of the best sites were used. The power would arrive in a predictable fashion for two four-hour periods each day. Electricity costs are slightly higher than nuclear or coal. Their future development will obviously have to take account of potential rises in sea level, and could be inhibited by uncertainties in this direction. Wave power, which is exploitable at some coastal sites as well as in the open sea, is at a relatively early stage of development. Two prototypes in Norway have been producing economic electricity since 1986, though one has been badly damaged in a storm.

Solar power, the most abundant source of energy, can be exploited actively by devices which convert it into electricity or hot matter, or passively by designing buildings in such a way that they absorb and retain the sun's heat. The former method is more readily available to countries with a lot of sunshine and a lot of space. The latter can make a valuable contribution in cooler countries, and is already being encouraged by the UK Department of Energy, through architects introducing large south-facing windows, conservatories and atriums.

Biomass energy is available from two sources. The first is organic waste matter of various types – domestic and industrial refuse, sewage sludge, farm slurry, and the unwanted residues of crops and forestry. These can be incinerated for heat or electricity or biochemically converted to fuels. The gases – principally methane – generated by decomposition in refuse sites can also be harnessed. Methane is converted to carbon-dioxide, a less potent greenhouse gas, in the process. The second source is crops grown for direct combustion or fuel conversion. Burning biomass creates carbon dioxide and other pollutants but these disadvantages can be offset by environmental gains. Waste that is not converted to energy is an expensive liability which also carries the risk of pollution. The carbon dioxide generated by biomass combustion is reabsorbed as long as the fuel crops continue to be grown, so there is no net addition of the gas to the atmosphere.

Geothermal energy is not properly speaking a 'renewable', but where it is available the source is virtually limitless. There are two types: underground reservoirs of hot water which can be directly exploited, and deeper-lying hot rocks which can be tapped by pumping water down for heating. There is a small hot water scheme in the UK, at Southampton, while hot rock sources are being explored by the Cambourne School of Mines in Cornwall.

Mike Flood, an independent energy researcher based in Milton Keynes, estimates that renewables could meet 20 per cent of Britain's energy needs by the year 2020, if energy efficiency measures to reduce Britain's energy demand are taken. 'The problems,' he says, 'include the intermittent or variable nature of some of the sources, high costs associated with pilot schemes, and lack of official support.' The Department of Energy, assuming a growth in energy consumption, puts the contribution at 8 to 10 per cent. The role of renewables in tackling global warming is discussed further in chapter 7.

Soaking up the carbon

There is an alternative approach to reducing carbon dioxide concentrations, which is to remove the gas from the atmosphere by the natural process of photosynthesis. One suggestion, which we noted earlier, is to stimulate algae growth in the oceans. The only practical course, however, is to grow trees – the one form of terrestial vegetation capable of absorbing and storing sufficiently large quantities of carbon. It is a partial solution which has a number of enthusiastic supporters. Gregg Marland, of the Oak Ridge National Laboratory in Tennessee, who researched the issue for the US Department of Energy, concluded that it was physically possible to remove 5 billion tonnes of carbon annually from the atmosphere by planting fast growing forests over an area equivalent to the total amount of forest cleared to date. An alternative to planting new forests would be to double the productivity of existing forests by restocking and harvesting the timber.

It would be a massive task, whichever way it was approached, and there are several practical objections. There is presently no foreseeable economic justification for planting forests on this scale. The adoption of bio-mass fuels, industrial feedstocks, the expanded use of timber and timber products might go some way to re-moving this difficulty. If the forests were used as biomass they would not in themselves produce a net reduction of atmospheric carbon dioxide unless they replaced fossil fuel consumption.

For new forests, there is the problem of finding some-where to plant. 'The competition for land resources for agriculture, biomass fuels, and human infrastructures is very great indeed,' Marland comments. It is this compe-tition, indeed, which has led to the continuing destruc-tion of forests. The land might be easiest found in the mid-latitudes if agricultural technology continues to pro-duce more food than is needed in spite of any climate change. Planting new woods and forests is already a preferred option for removing land from agricultural use in the UK. The contribution Britain could make to ameliorating climate change would be marginal, however. An area the size of Western Europe would have to be planted to take up 15 to 20 per cent of current man-made carbon emissions. There is also the question of whether forests would flourish in these regions under the combined impact of climate change and existing pollution stresses. Forests in East and West Europe and North America are dying. Those in the south-east USA are growing more slowly – by 30 to 50 per cent – than they did thirty years ago.

An ambitious afforestation programme aimed at mak-ing a significant difference to climate change would be pointless if the destruction of tropical forests were to continue at its present accelerating rate. It would at best represent a case of two steps forward for every one back. The case for conserving the tropical forests as reservoirs of bio-diversity and regulators of local climate stands on its own merits, regardless of global climate change. It is a case, however, which appears less evident to the tropical forest countries, hungry as they are for land and export earnings. It will require skilful diplomacy and a new approach to international economics on the part of the developed nations to win them round. The problem was plainly stated by David Everest in his Joint Energy Programme report. The initial rise in atmospheric carbon dioxide, he pointed out, was due to forest clearance in countries such as the USA.

It would be politically difficult for the economically advanced countries, which have in the past made their own contribution to such CO_2 emissions, to attempt to stop similar development in less developed countries. Although environmentally beneficial in its own right, this policy component does not appear to represent significant means of limiting the onset of the greenhouse effect, or even causing its reversal, as has sometimes been suggested.

Everest may have had in mind the statement of Adam Malik, Vice President of Indonesia, who told the 1982 Third World National Parks Conference in Bali:

How much land for the hungry of today? And how much for genetic resources to be preserved for tomorrow? In the past, we have neither received a fair share of the benefits, nor have we received a fair share of assistance – other than inexpensive advice and even more inexpensive criticism – in the efforts to save the common global natural heritage. Unless such responsibilities are equally shared, all our good intentions will only lead to global environmental destruction.

Setting the agenda

A substantial body of serious world opinion is sufficiently convinced by the threat of global warming to call for immediate action to slow down climate change and cope with its consequences. In Toronto, in June 1988, more than 300 experts from forty-six countries agreed that the concerns about atmospheric change were justified and that 'the time to act on the problem is now'. They included the prime ministers of Canada and Norway, ministers from other countries, leaders in science, law and the environment, officials from international agencies and non-governmental organisations.

Four months later the island state of Malta presented to the UN General Assembly a successful resolution calling on governments, intergovernmental and non-governmental organisations to treat the problem of climate change as a priority and to collaborate in making every effort to prevent further detrimental effects. Britain was one of the co-sponsors. A fortnight later, in Hamburg, a World Congress on Climate and Development declared, 'There is an urgent need for political and scientific leadership, at the highest level, as well as a need for industrial and consumer action, to reduce emissions of greenhouse gases.'

Almost simultaneously, in Geneva, an Intergovernmental Panel on Climatic Change (IPCC) from thirty-five nations, convened by the UNEP and the WMO, set up three expert committees to review the scientific evidence, assess the environmental, economic and social impacts, and devise response strategies. This was the first practical step towards an international convention on climatic change, to be debated and endorsed by world leaders in 1992. Mostafa Tolba, head of the UNEP, said the conclu-

Insurance Policy

"If I don't believe in God, and he exists,
- then I'm in big trouble.

But if I do believe in God, and he doesn't exist,
- then it's no big deal." *After Pascal*

Fig. 12

sions of the expert committees should be sufficiently advanced by mid-1990 to justify actions by governments and start negotiations for a treaty.

A few days later, the European Commission laid before ministers a draft resolution welcoming talks on an atmosphere protection treaty and setting out proposals for action.

The Toronto statement declared:

> The Earth's atmosphere is being changed at an unprecedented rate by pollutants resulting from human activities, inefficient and wasteful fossil fuel use and the effects of rapid population growth in many regions. These changes represent a major threat to international security and are already having harmful consequences over many parts of the globe . . . The best predictions available indicate potentially severe economic and social dislocation for present and future generations, which will worsen international tensions and increase risk of conflicts between and within nations. It is imperative to act now.

The conference called for a 20 per cent reduction in carbon dioxide emissions by the year 2005 'as an initial global goal', including a 10 per cent improvement in energy efficiency. This was the first time a major gathering of this type had called for target reductions.

The Hamburg Congress called for action by nations to reduce carbon dioxide emissions globally by 30 per cent by the year 2000 and 50 per cent by 2015. It argued for unilateral action from the industrialised nations to start the process of change; a global ban on the production and use of CFCs covered in the Montreal Protocol by 1995, with other ozone-depleting substances added as scientific evidence dictates; and urgent strategies for reversing deforestation and beginning afforestation programmes.

The UNEP hopes to see a framework Convention on

Making it work

How can policies to reduce the output of carbon dioxide and improve energy efficiency be made to work? Controls already exist on the emissions from power-stations and other large furnaces for air pollutants like smoke, grit and sulphur dioxide. They depend, however, on the existence of practicable technologies for removing the pollutants from the chimney and disposing of them safely. Carbon dioxide is not classed as a pollutant and there is no economically viable technology for removing and disposing of it. The same problem applies to any attempt to control emissions from motor vehicles. Possibly the only feasible way to control carbon dioxide with existing pollution control methods is by enforcing the use of cleaner fuels. Other measures, such as taxation, designed to reduce fuel and energy consumption, might be more practicable.

Setting efficiency standards for electricity users is somewhat easier. In the UK, building regulations already specify standards of insulation. Obligatory standards could be applied to electrical motors and appliances, heating, lighting and ventilation systems, and industrial processes.

Another potential method of control is the town and country planning system. Under a recent EEC directive, all applications for planning permission to carry out major development – including power-stations, oil refineries, chemical and steel works, and major roads, railways and airports – must be accompanied by an **environmental impact assessment** (EIA). This is a document describing all the environmental damage and disturbance likely to be caused by the development and the developers' proposals for minimising or otherwise dealing with them. Among the issues it must cover are the noise, and extra traffic involved in construction, the ecological or agricultural value of the land taken by the development, and the pollutants flowing from the plant. The impact on climate is specifically included. The developer must show he has considered other ways of meeting his objectives. In the EIA for the Hinckley C nuclear power-station, for example, the CEGB examined the possibility of generating the power by other means.

Submission of an EIA is obligatory for specified major developments and may be required by the local planning authority for other developments. The device has been used voluntarily on more than 200 schemes in the UK, but there is as yet little experience of its statutory use. As with most aspects of planning law in the UK, it is not a matter of rigid regulation.

County Structure Plans in Britain are another mechanism for achieving control of greenhouse gas emissions. Local authorities can already specify energy efficiency standards in all new houses built in their area, as Milton Keynes have done with their Energy Conservation Index (ECI) (see chapter 7). They can also positively encourage renewable energy sources, combined heat and power-stations. Both Structure Plans and EIAs are potentially tools of immense power if used in conjunction with strong planning policies set at the national and local level. It is clearly within the scope of the EIA regulations to say that a coal-fired power-station should not be built because the same amount of electricity could be produced by less-damaging means.

The EIA could be used to ensure that a new town plan embodied the highest attainable standards of energy efficiency. It could also be used to consider whether the scheme reduced the need for commuting and private car journeys by providing jobs and facilities close to homes. It could assess proposals for a new motorway against the provision of public transport.

Local planning authorities have the right to do deals with developers for what is called **planning gain**. Usually, the developer agrees to build, fund, or provide land for public roads and other general amenities.

In the context of climate change, the deal could include planting trees to soak up carbon dioxide, insulating existing houses to reduce carbon dioxide emissions, or providing a sustainable source of biomass energy.

Climate Change, modelled on the Vienna Convention on Stratospheric Ozone, negotiated and agreed by 1992. This should be an agreement to stabilise greenhouse gases in the atmosphere at a level that will keep climate change to a minimum, and its impacts within a range that can reasonably be managed. It should encourage developing countries to play their part in the global effort by the provision of financial and other incentives, including the transfer of technology and expertise from the developed nations. The convention should also cover scientific, technical, legal and policy co-operation, monitoring arrangements, and exchanges of information.

Detailed agreements would take the form of 'protocols', akin to the Montreal Protocol on CFCs. To avoid delays, UNEP would like to see the first protocols agreed at the same time as the convention. Protocols should cover targets for reducing emissions. A 20 per cent cut in carbon dioxide by the year 2000, achieved in equal measure by energy saving and changing fuel sources, and a long-range cut of 50 per cent, have been suggested. No target dates or volumes have been suggested yet for methane and nitrous oxide. It is proposed that countries should be left to adopt the most appropriate means of control, including controlling emissions from cattle feedlots, altering agricultural practices, preventing gas leaks, and controlling vehicle emissions.

Can there be any doubt that we must act as soon as possible? The onset of global warming is not yet proven, but the evidence is strong. Many of the things we need to do to reduce the greenhouse effect will bring us other benefits. The economic case for energy saving is unshakable. Reductions in the pollutants of fossil fuel combustion will benefit the environment and our health. Saving the forests will secure the potential benefits of countless species. We have so little to lose, and so much to gain.

6

NUCLEAR POWER — SAVIOUR OR FALSE PROPHET?

'If the situation is as serious as the Toronto Conference depicts it,' declared Hans Blix, Director-General of the International Atomic Energy Agency (IAEA), 'the time would seem ripe for the "greens", for those who favour the use of nuclear power and for others, who are unsure where they stand, to discuss without acrimony what practical measures can be taken to avert disaster.' Blix was in no doubt what the conclusions of such a meeting would be. 'Nuclear power is the only source that is now available to generate electricity in the quantities, form and reliability needed without producing any of the greenhouse gases.'

Blix's audience were delegates to the thirty-second session of the IAEA in Vienna in September 1988. They were reportedly enthusiastic about Blix's theme that those who supported nuclear power as an option to reduce the greenhouse effect were the true 'friends of the earth'. It was a rallying call and a theme to be taken up by many others in the following months.

The theme was developed by British Prime Minister Margaret Thatcher in an interview with *The Times* in October 1988:

Some of the people who are critical of the environment aren't exactly helping to get more nuclear power —

which, of course, would deal both with the green-house effect and acid rain . . . We have to be looking at having a much heavier nuclear programme.

Mrs Thatcher's Environment Secretary, Nicholas Ridley, strongly supports her nuclear views. 'The nuclear pro-gramme is the only serious way of reducing enough carbon emissions,' he told a TV reporter. 'We should concentrate like the French are on a massive increase in nuclear generating capacity.' In similar vein, a Tory member of the European Parliament, Richard Cottrell, warned his constituents: 'the seasons are changing. There won't be any more white Christmases . . . Britain should phase out coal and oil power-stations and go all out for nuclear energy instead.'

In Canada, Jovan Jovanovitch, Professor of Physics at the University of Manitoba, is an unashamed advocate of the so-called 'super hard' energy path – namely, a very fast growth in the use of nuclear power. In his view, nuclear power 'is the only energy strategy that can create an industrialised and affluent world in a relatively short time, and without adverse impact on the environment' and regards it as the only 'clean' energy source currently available.

In the US, President Bush has been less forthright in his support for nuclear power, mainly due to the poor economics and safety record of nuclear energy in his country. Support for nuclear power brings few votes. At a news conference prior to his inauguration as president, however, he expressed support for 'safer, friendlier nu-clear reactors as one of the options to solve the green-house effect.'

In France, Jean-Pierre Capron, general administrator of the Atomic Energy Commission (CEA), is 'convinced that the key argument in favour of nuclear power is the protection of the environment' and that it is the 'only way to afford growth without triggering a major ecological disaster.' Capron thinks that the contribution of nuclear energy in a world which he assumes will be using 250 per cent more energy by the middle of the next century, 'could well be a decisive one'. Meanwhile, French Environment Minister, Brice Lalonde, a former presidential candidate for the Green Party, was reconsidering his opposition to nuclear power and told reporters he was 'reassessing the whole thing' in the light of global warming. France is of course the most pro-nuclear country in the world and depends on nuclear power for 70 per cent of its electricity and over 20 per cent of its total energy.

Nuclear power is, along with a range of renewable energy sources such as the sun, wind and biomass, one of the options frequently suggested as a replacement for the combustion of carbon-emitting fossil fuels. At first glance, the expansion of nuclear power seems an attractive choice. Indeed, as we have seen some politicians and scientists have been quick to nail their colours to the nuclear mast. Behind the obvious propaganda use of global warming by nuclear supporters to reduce public opposition and hence expand their industry, what sort of proposition is being put forward? How extensive a nuclear programme would be needed? How effective could nuclear power be in reducing greenhouse gas emissions? How much would such a programme cost?

To avoid any well-rehearsed pro- and anti-nuclear arguments, we will assume something which the nuclear industry could only dream of – a new beginning. Any concern and uncertainty of radioactive waste, reactor accidents, and the spread of nuclear weapons through proliferation, will be put to one side. We will also assume that nuclear power-stations can be built to time and budget (something which has not been the case in many countries), that they operate efficiently, and that public opinion has swung in favour of the nuclear option. Let us now examine the potential role of nuclear power in reducing global warming, recalling the Bellagio workshop conclusion that 50 per cent reductions in carbon

dioxide emissions are necessary to keep global warming
rates down to 0.1 °C a decade (see appendix 1).

A billion a day

The starting point for nuclear power is relatively low. It
provides around 3 per cent of the global delivered
energy, in the form of electricity. However, most of the
world's energy is not provided by electricity: it consists of
liquid fuels such as petrol and diesel in vehicles, oil used
in chemical plants, coal and gas in domestic central
heating and industrial boilers, and wood burned in cooking stoves. Although nuclear power-station capacity has
increased rapidly since 1975, world energy consumption
has also grown – more cars are in use, a larger population
is burning more and more wood and coal, and industry is
producing more chemicals from oil. Nuclear power's
share of the world energy pie thus remains a small one.

Energy demand used to be thought of as closely linked
to economic and population growth. This is no longer the
case. Energy efficiency allowed Japan to increase its GDP
over ten years by 46 per cent while reducing energy
consumption by 6 per cent. The pattern is true of many
other industrialised countries. To measure the potential
role of nuclear power, we have to make some assumptions about the overall demand for energy in the future:
will it reduce, stay the same, or go up – and if so, by how
much? Energy researchers do this by producing what are
known as 'scenarios', which are not forecasts but a
plausible evolution of energy use based on assumptions
of economic growth, technology uptake, rates of energy
efficiency, the numbers of vehicles and so on. In each
scenario you can then work out how many nuclear
power-stations or coal mines you would have to con-
struct to provide a certain proportion of the total demand, or to achieve target cuts in carbon dioxide
emissions.

Increases in nuclear capacity figure prominently in many government and electricity-industry energy forecasts. Despite this, the nuclear industry has been remarkably coy about what size of increase it is now advocating to reduce global warming. One of the few nuclear advocates to specify target increases is Alvin Weinberg. Weinberg, a physicist who worked on the Manhattan Bomb Project and the pressurised water reactor (PWR), argues that carbon dioxide provides 'the strongest incentive to get back on track with nuclear energy'. He called for a six-fold increase in current capacity at a Los Alamos conference in 1987.

Gregory Kats and Bill Keepin, from the Rocky Mountain Institute in Colorado, looked at three scenarios recently, in a detailed analysis of the potential roles of nuclear power and energy efficiency in reducing carbon dioxide emissions. They also assumed that nuclear power had resolved all its economic and safety problems. The first two energy scenarios project increases in energy consumption ranging from 210 to 360 per cent by the year 2020. They are based on work by groups such as the World Energy Conference (WEC), the US Department of Energy, the US National Academy of Sciences, and the International Institute for Applied Systems Analysis (IIASA). The third is a 'low energy scenario' which projects only a 10 per cent increase in 2020 from current levels.

Keepin and Kats set out to calculate how many nuclear reactors would be required to displace coal entirely from the energy mix over a period of a few decades. The results were dramatic. In the high-energy growth scenario – a 360 per cent increase from today's levels – more than 8,000 large nuclear plants would have to be built over the next thirty-seven years, at the rate of one every day and a half. The capital cost would be $8.39 trillion (million million), or an average $227 billion each year. This is double the annual industrial investment capital of the USA, and twenty times the total cost of the Channel Tunnel, every year (see fig. 13).

Although nuclear power manages to displace all coal-fired power-stations in the electricity sector, overall carbon dioxide emissions still increase by 65 per cent from the 1987 levels. Fossil fuel use in the transport, domestic and industrial sectors has increased and simply swallowed up the carbon dioxide savings from the nuclear reactors. This gargantuan nuclear programme, which would dwarf any engineering project ever undertaken, including the Manhattan Bomb Project, has managed to slow down the rate of increase in carbon dioxide emissions, but failed by a large margin to actually reduce them. In the light of the calculations, it is difficult to see how Jean-Pierre Capron of the CEA can argue that the contribution of nuclear 'could well be a decisive one' in averting 'a major ecological disaster'.

The results for the middle energy growth scenario – a 210 per cent increase – are little better. In this case, a nuclear plant must be built every three days at a total capital cost of $5.3 trillion, or $144 billion per year. The net result of having 5,000 large nuclear plants operating by 2020 is still a 10 per cent increase in carbon dioxide emissions. Moreover, 40 per cent of the reactors would have to be in less developed countries (see 'How much carbon dioxide can nuclear power save?', pp. 171–172). Keepin and Kats conclude: 'any global energy scenario which assumes modest or rapid growth in energy demand leads to increased carbon dioxide emissions that no conceivable nuclear programme could alleviate.'

Few, apart from supporters of the 'super hard' energy path such as Jovan Jovanovitch, would seriously believe that the seventeen- to twenty-seven-fold increases in nuclear capacity assumed in the scenarios are feasible. Keepin and Kats therefore assessed one additional scenario, a six-fold increase in nuclear capacity – the figure suggested by Alvin Weinberg. This scenario entails a more modest rate of nuclear construction – a nuclear

Nuclear Power - Solution to Global Warming?

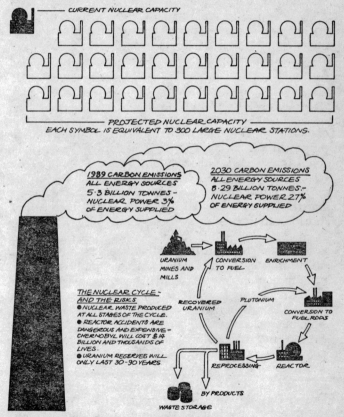

A NUCLEAR SCENARIO
UNLESS GLOBAL ENERGY DEMAND IS HELD STEADY THROUGH ENERGY EFFICIENCY, EVEN BUILDING A NUCLEAR PLANT EVERY 1·6 – 3 DAYS AT A COST OF $5·3 – $8·4 TRILLION, CARBON DIOXIDE EMISSIONS COULD STILL INCREASE 65% OVER 40 YEARS *

— CURRENT NUCLEAR CAPACITY

— PROJECTED NUCLEAR CAPACITY —
EACH SYMBOL IS EQUIVALENT TO 300 LARGE NUCLEAR STATIONS.

1989 CARBON EMISSIONS
ALL ENERGY SOURCES
5·3 BILLION TONNES –
NUCLEAR POWER 3%
OF ENERGY SUPPLIED

2030 CARBON EMISSIONS
ALL ENERGY SOURCES
8·29 BILLION TONNES –
NUCLEAR POWER 27%
OF ENERGY SUPPLIED

URANIUM MINES AND MILLS — CONVERSION TO FUEL — ENRICHMENT

RECOVERED URANIUM — PLUTONIUM — CONVERSION TO FUEL RODS

THE NUCLEAR CYCLE – AND THE RISKS
● NUCLEAR WASTE PRODUCED AT ALL STAGES OF THE CYCLE.
● REACTOR ACCIDENTS ARE DANGEROUS AND EXPENSIVE – CHERNOBYL WILL COST $14 BILLION AND THOUSANDS OF LIVES.
● URANIUM RESERVES WILL ONLY LAST 30 – 90 YEARS

REPROCESSING — REACTOR

BY PRODUCTS
WASTE STORAGE

* AFTER KEEPIN AND KATS, ROCKY MOUNTAIN INSTITUTE, 1988

Fig. 13

plant every 7.5 days, at a cost of $49 billion per year (or four Channel Tunnels). Under the high and medium energy growth assumptions, this nuclear programme barely dents increases in carbon dioxide emissions at all. It once again fails to move the world closer to a target goal of 50 per cent carbon dioxide reductions, and a warming rate of 0.1°C per decade.

What is true for nuclear power in a period of rising energy demand is, of course, also true for other sources of electricity designed to replace fossil fuels. So do any of these options have a role to play in combating global warming? The answer is yes, but only as part of a co-ordinated strategy to tackle the problem which has energy efficiency firmly at its heart.

Keepin and Kats carried out a comparative assessment of the cost-effectiveness of energy efficiency and nuclear power in displacing carbon emissions. Their results show that energy efficiency is seven times more cost-effective. Energy efficiency measures to achieve carbon dioxide reductions are outlined in chapter 7. Their work and that by other researchers (see 'How much carbon dioxide can nuclear power save?', pp. 171–172) isn't an attempt to suggest that any of these massive nuclear programmes are realistic. Indeed the sheer cost ensures that no country could afford them. They simply show that putting all our eggs, or indeed even a sizeable number of eggs, in the nuclear basket isn't going to solve global warming. It demonstrates vividly that a massive expansion in nuclear power capacity can, in the words of Keepin and Kats: 'only take a small bite out of the greenhouse warming pie – unless the pie itself is pre-shrunk by improved energy efficiency.'

One energy scenario which does take a hefty bite out of the greenhouse warming pie is the 'low-energy scenario' developed by the Brazilian economist Jose Goldemberg and his colleagues Thomas Johannsen from Sweden, Robert Williams from the US, and Amulya Reddy from

How much carbon dioxide can nuclear power save?

Electricity produces around 35 per cent of the world's carbon dioxide emissions, though only 15 per cent of the world's energy, and nuclear power use currently provides around 3 per cent of world energy end-use needs. Several studies into the potential role of nuclear power in displacing fossil fuels and reducing carbon dioxide emissions have shown that:

☐ For nuclear to replace 50 per cent of the world's current fossil fuel energy use, amid 3 per cent economic growth, it would require building sixteen large reactors per week between 1995 and 2020. This is fifty times the peak rate of construction between 1975–1985. To accomplish the same goal with improved energy efficiency would require a 4.6 per cent per year improvement in the world's gross national product/energy ratio. This is about 60 per cent greater than the US achieved during 1976–86, 35 per cent faster than the rate in Japan, and double that in Britain.

☐ If world energy demand doubles or trebles in the next thirty years, as some forecasters predict, carbon dioxide emissions would rise between 10 and 65 per cent, even if a nuclear reactor was built every one and a half to three days. This rate of construction would increase the world nuclear capacity from 300,000 megawatts by between 5,000,000 and 8,000,000 megawatts in the year 2020, and cost from $5.3 to $8.39 trillion.

☐ In England and Wales, the construction of twenty-five large nuclear reactors between 1990 and 2030, at a cost of $45 billion, would fail to reduce carbon dioxide emissions if demand rises as the utility, the CEGB, predicts. Emissions would rise by 20 per cent (8 per cent of total carbon dioxide emissions) in England and Wales.

□ In Eastern Europe, a crash nuclear building programme to increase capacity to 70 per cent of total
electricity supply in order to cut overall carbon
dioxide emissions by 25 per cent, would require
building twenty-nine nuclear plants every year.

India. Their scenario, called *Energy for a Sustainable World*,
projects a mere 10 per cent increase in global energy
demand from today's levels by the year 2020. It is one of
twenty-seven 'low-energy scenarios' which have now
been developed for fourteen individual countries as well
as the whole world in the past decade. Keepin and Kats
calculate that a six-fold increase in nuclear capacity under
this scenario, linked to a policy of using more natural gas
in place of coal and greater emphasis upon biomass,
would actually cut carbon dioxide emissions by 38 per
cent. The message is clear: nuclear advocates must therefore look to energy efficiency for their expansion plans to
have any meaning.

Even a six-fold nuclear expansion comes with a hefty
price tag – $1.8 trillion. It would also indirectly generate
its own carbon dioxide emissions. Fossil fuels have to be
used in mining and enriching uranium ore, producing
the cement and metal for the power-stations, dealing
with the waste products, and decommissioning old nuclear reactors at the end of their short thirty year lives.
Nigel Mortimer, an energy analyst from Sheffield Polytechnic, calculates a figure of 250,000 tonnes of carbon
dioxide per year over the thirty year lifetime of a large
PWR. This would increase once the richer uranium ores
are used up. The construction of large numbers of nuclear reactors would thus initially make the carbon dioxide problem worse. Every $100 invested in new nuclear
power capacity adds approximately one tonne of addi-

tional carbon into the earth's atmosphere that could have been avoided, had the money been used instead for improved efficiency.

Time for environmentalists to think again?

Should environmentalists take Hans Blix's advice, reconsider their opposition to nuclear power, and sit down with the nuclear industry? Although in theory nuclear power could have a role in combating global warming if we are very successful at reducing energy consumption, there are other environmental and safety implications to consider. In 1987, the World Commission on Environment and Development, chaired by Norwegian Prime Minister Gro Harlem Brundtland, produced its long-awaited report, *Our Common Future*. (See 'The Brundtland Report', pp. 94–95). The Commission took evidence on energy problems and proposed solutions throughout the world. It received extensive material from all sides of the nuclear debate. Discussing the issue of nuclear power in the context of the world's energy needs, the Commission concluded:

> The generation of nuclear power is only justifiable if there are solid solutions to the presently unsolved problems to which it gives rise. The highest priority must be put into research and development of environmentally sound and economically viable alternatives as well as of means of increasing the safety of nuclear energy.

The unsolved problems the report highlighted were the risk of catastrophic accidents, the management of high-level radioactive waste, and the proliferation of nuclear weapons.

Any attempt to assess both the extent of these problems, and possible means of ameliorating them, is

A short history of nuclear power

The first sustained nuclear chain reaction had a humble location in a disused squash court in Chicago in 1942. In 1945, the awesome power of the atom was first used in anger during the Hiroshima and Nagasaki weapons explosions.

Nuclear power for the generation of electricity grew out of weapons programmes in the USA, Britain, France and USSR. The world's first civil reactors started operating in the UK and USSR in 1957. President Eisenhower opened the doors to the widespread use of nuclear power with his famous 'Atoms for Peace' speech to the United Nations in 1957, which led to the training of thousands of nuclear scientists and technicians, as well as the spread of fissile materials for research and commercial power production.

A wide variety of reactor types have been developed in the past thirty-five years. Britain originally opted for gas-cooled reactors – the **Magnox** and **Advanced Gas Cooled reactors** (AGRs) – though she was later to choose the world's most popular design, the **Pressurised Water Reactor** (PWR). The PWR has captured 62 per cent of the reactor market, followed by the **Boiling Water Reactor** (23 per cent), the **CANDU** (5 per cent), and **Gas-Cooled Reactors** (4 per cent). The PWR was originally a compact design for a submarine reactor, developed under the direction of Captain Hyman Rickover.

The plutonium-fuelled **Fast Breeder Reactor** (FBR) is cooled by liquid sodium, and has always been viewed as a natural successor to the other types of reactors when uranium ore becomes scarce or expensive. Of the nineteen experimental, demonstration and 'commercial' FBRs so far built in the past twenty years, only three are regularly providing electricity to the grid, of which only one could be regarded as commercial.

Nuclear power was heavily subsidised in its early days, and still is in many countries such as Britain, the USA and France, where it is dependent on government support. Britain is instigating a 'nuclear tax' to protect the industry when it is privatised in 1990, a tacit admission that nuclear is more expensive than gas and coal stations. It is also supported by a large number of institutions such as the International Atomic Energy Agency (IAEA), Euratom, and the Uranium Institute.

The IAEA was formed by the United Nations in 1958, to control the movement of fissile material and prevent its possible diversion for weapons purposes, as well as to promote peaceful uses of the atom. Twelve countries are known to either have nuclear weapons or the capability of making them. India exploded a weapon in 1974, using civil nuclear technology.

There have been twelve accidents at reactors involving loss-of-coolant (LOCA) and serious core damage. Most have involved an element of human error. The most serious was at Chernobyl in the Ukraine in 1986. Thirty-two people have died so far, and an estimated 5,000–20,000 long-term fatal cancers may result. The total cost of the accident is over £10 billion.

A series of reactor cancellations from 1973 onwards has slowed the US nuclear industry virtually to a standstill. The utility chairman of Tennessee Valley Authority has described it as an option 'no sane investor would touch'. France and Japan have now taken over from the USA as the major nuclear countries. Nuclear power provides over 70 per cent of France's and one-third of Japan's electricity. France now exports excess electricity to six other countries in Europe, but is suffering from a serious debt problem to the tune of £23 billion, due to building many more reactors than are needed.

fraught with difficulties. Although there are certain *facts* relating to each problem, which are generally agreed on by both sides of the nuclear debate, there are also a wide range of grey areas where *judgement* is required as to whether solutions are available or not. Take the hot issue of radioactive waste.

Nuclear waste – a problem for our children

The development of nuclear power is based on a faith which the public place in technology to solve all problems.
        Dr Mike Pasquellati, Arizona State University, 1988.

In the first three decades of nuclear power, little thought was given to the disposal of radioactive waste. It was regarded as an insignificant problem which could be left to a future date. That future date has now arrived. Stockpiles of waste at power-stations and reprocessing plants urgently need disposal or storage solutions.

In terms of volume, the amount of waste produced from all stages of the nuclear-fuel cycle – uranium mining, enrichment, fuel fabrication, reactor operation and reprocessing – is relatively small. For example, it is an order of magnitude lower than industrial chemical wastes and several orders of magnitude lower than domestic waste. What makes radioactive waste so special and gives rise to so much public concern is the radioactivity contaminating it.

Radioactivity decays over time, emitting various types of radiation as it does so. Although you cannot see, taste or smell it, human contact with radiation can lead to damage of the DNA structure of cells, which in turn can cause cancers and possible death. We are exposed to it in a variety of ways, such as cosmic rays from outer space, radon gas from certain rocks (including granite), and medical X-rays. It is estimated that over 100,000 people die annually from cancer worldwide as a result of so-called 'natural' radiation and medical exposure. The question we have to address is how much extra risk from man-made radioactivity entering the environment from the nuclear fuel cycle is acceptable?

The long-term objective for radioactive waste management is to ensure that radioactive pollution does not reach the human environment, or does so at such a point in the future that the amounts of radioactivity involved are fairly harmless. Much of the recent nuclear controversy in countries such as the USA, UK, Poland and Spain has been over proposals for nuclear waste dumps – or 'repositories' as the industry prefer to call them. Although effectively banned since 1983, the sea dumping of radioactive waste has also caused major international unrest and concern. Repositories can be either shallow burial sites, perhaps only twenty to thirty metres beneath the surface, or deeper burial sites on land, below the sea-bed, or in tunnels which start on land and then extend under the sea-bed.

Only a small number of waste repositories for civil nuclear waste have so far been built. The operating record is mixed. Two of the six commercial sites in the USA have been closed after a number of leaks into the environment. In France, one well-engineered site has been operating for over sixteen years with no leaks reported to date. In the UK, a repository at Drigg near the Sellafield reprocessing complex has been slowly leaking low-level waste into the Irish Sea for over twenty years as it was in fact designed to do. In the Soviet Union, a waste dump exploded at Kyshtm in the Urals in 1957, contaminating hundreds of square kilometres of land and many thousands of people.

The dumps developed so far are designed mainly for low and so-called short-lived intermediate level waste – wastes which become relatively harmless after several hundred years. It is the longer-lived wastes, including the high-level or 'heat-generating' wastes from spent nuclear fuel, which cause the greatest concern.

High-level wastes include very long-lived isotopes which remain dangerous for thousands of years. Scientists try to isolate them from the human environment through a whole series of man-made and natural geological barriers. It is extremely difficult to convince a sceptical public that this can be done over the long time-scales involved. A former UK Environment Minister, William Waldegrave, admitted:

> The consequences cannot be empirically judged for some 200 years or so. Thus there is no way of being able to demonstrate to the public that you have done it and that it succeeds. So by its nature, in an area of this kind, you have to proceed by trust.

It is this trust which has been clearly lacking in the exchanges between the nuclear industry and the public.

Can high-level waste thus be disposed of safely? The nuclear industry and its supporters are confident that it can. Hans Blix of the IAEA says 'an international consensus exists on principles for the safe disposal of high-level waste', a conclusion with which the House of Lords European Communities Committee recently concurred. However, agreement on the principles does not constitute proof that it can actually be done.

From 1984 to 1987 the Beijer Institute of the Royal Swedish Academy of Scientists sponsored a detailed study in the disposal of high-level radioactive waste. The team of researchers carried out a comparative analysis of proposed disposal programmes in eight major nuclear countries with 75 per cent of world nuclear capacity. Although the researchers noted considerable progress in research on geology, the behaviour of radioactive materials over time, and waste packaging, they found it impossible to conclude that a solution to the problem of high-level waste disposal had been found. Their report said:

> The solution to the high-level radioactive waste problem is not and cannot be a totally technological problem . . . it is an inter-disciplinary problem that requires legal, social, political and technical input. Proving a solution to the long-term and high-level radioactive waste problem poses difficulties of a kind that have never been addressed before in scientific and engineering investigations, and to some extent are more properly within the realm of philosophy.

The main problem is to prove unequivocally and to the satisfaction of the public that scientists have foreseen all the possible problems that might arise over the long timescales involved. These problems include human interference with the repository, seismic disturbance, and rapid leeching of the radioactivity through faults and rock fractures. Any repository design must also take account of large changes in sea levels due to global warming or another ice age, when sea levels can fall several hundred metres.

Has the high-level waste problem been solved? The answer is 'not yet', but some scientists are optimistic and confident of technological solutions. Others are less sure. Accidents can and do happen.

Reactor accidents

> It is therefore fair to explain to the public that we do not expect big reactor accidents to happen, the chance of them happening is remote, and that even if one did happen, then on average their life expectancy would be reduced by this small amount (twenty hours) due to the adverse health effects of radiation released during the accident.
>
> Lord Marshall, Chairman of the CEGB, 1982.

The disaster at Chernobyl on 26 April 1986 was mainly the fault of operators, who conducted an unauthorised

Nuclear power worldwide – a status report

At the end of 1987 there were 417 nuclear reactors operating, with a capacity of 300,000 megawatts. They provided 16 per cent of the world's electricity, but only 3 per cent of total energy. Twenty-one countries have commercial nuclear programmes. Nuclear power is both a major source of electricity and strongly supported by the governments of only five or six countries: France, Japan, Belgium, South Korea, Taiwan and possibly the USSR. The latter has cancelled six reactors since the 1986 Chernobyl disaster.

The US, which operates more nuclear reactors than any other country, will have a declining nuclear capacity from the mid-1990s under current projections. A number of countries such as Britain, West Germany and Spain are replacing older reactors rather than adding to nuclear capacity.

There is a growing 'non-nuclear' group of countries, which have either decided not to use nuclear power, or are getting rid of it. They include Austria (whose completed reactor at Zwentendorf may be turned into an amusements park); Italy and Sweden (where referenda are leading to nuclear phase-outs); and Denmark, Ireland, Greece, and the Philippines.

In the developing world, India has ambitious plans to emulate Taiwan and South Korea, but may not complete a fifth of its 10,000 megawatts target by the end of the century. Construction delays, cost overruns, and technical problems have been the hallmark of the nuclear programmes in Argentina, Brazil and Mexico. China has cancelled a ten-reactor programme due to high costs.

The rate of ordering nuclear reactors has slowed down considerably, particularly since the 1979 Three Mile Island accident and Chernobyl in 1986. Five major reactor construction companies are chasing possible orders for just eight to ten plants over the next twelve to fifteen years. They have the capacity to build at least four times that number.

Nuclear reactors are currently saving the equivalent of just over 750 million tonnes of carbon dioxide emissions per year – around 4 per cent of total carbon dioxide emissions – by displacing fossil fuels.

experiment with the reactor. With some thirty-two deaths so far and thousands more expected over the next forty years, it was by far the most serious of the twelve accidents which have led to serious reactor core damage over the past thirty years. It caused panic, confusion and the banning of some food products throughout Europe. Nuclear accidents have been due to both technological and human failures, and interactions between the two. Nuclear power is admitted by the industry itself to be still going through a learning curve, so accidents are hardly surprising as we discover new aspects of the technology. But how frequently will they occur in future, and what are the consequences?

According to the IAEA and Britain's Central Electricity Generating Board (CEGB), the chances of major reactor accidents involving serious radioactive contamination are about one in a million per reactor per year. 'One in a million' sounds a reassuring figure, but what does it mean? The figure is based on detailed computer calculations which assume failure rates for the many thousands of components within a nuclear station. Engineering 'judgement' is used at a number of stages where lack of operating experience makes it difficult to input reliable data.

Is this accident risk acceptable? Is it even reliable? A robust debate on the issue occurred in the pages of *Nature* in 1986. It was initiated by Dr S. Islam from West Germany and Dr K. Lindgren of Sweden. They suggested that the official claim that the chances of a reactor accident would be less than one in a million per year had 'not been met in practice, and that the method of technical risk assessment used to calculate such probabilities must be replaced by risk assessments which use different operating experience'.

The academics' own figures were hardly reassuring. On the basis of accidents and near accidents that had already taken place, rather than optimistic hopes for future safety performance, they suggested that there was a '70 per cent probability that one accident could happen in the next 5.4 years' and a '95 per cent probability of having one accident every twenty years'.

We thus have two opposing views on the risks of catastrophic accidents, neither of which takes account of the increase in risks resulting from any large expansion in nuclear power. One is a technologically optimistic view which relies heavily on engineering judgement, 'fail-safe' operating devices which can automatically shut down a reactor, and a belief that no nasty surprises are in store for us with the current designs of reactors. The other is a more pessimistic – some would say, pragmatic view – which assumes that mistakes occur, that technology regularly fails, and that the safety record of extremely complex technologies doesn't always improve through time. The public has to decide which of the two viewpoints to believe.

Perhaps the final word on reactor safety should be left to the nuclear industry itself. Morris Rosen, Director of Nuclear Safety at the IAEA, was asked by journalists in the aftermath of Chernobyl how many reactor accidents we could expect in the future. His answer was remarkably frank:

I might say to you that there will be an accident every ten years, or I might say to you once every five years or

even every year. You will have to tell me whether this
is acceptable or not . . . but what I can tell you is that
nuclear power would still be safer than any other
energy alternative.

In other words, Rosen is no longer arguing that there
won't be further catastrophic accidents in future, but that
a major accident once every ten or even five years might
be acceptable because, in his view, other energy sources
such as coal are more risky due to air pollution and other
health risks. Whether cross-boundary radioactive pollution is going to be politically and environmentally acceptable as a consequence of an attempt to reduce the
greenhouse effect is debatable. Another major accident
would provide the answer. It is more likely to kill off any
chances that nuclear power may have of becoming a
major energy source.

Nuclear weapons proliferation

For nuclear power to have a long-term role in providing
electricity and thereby reducing the world's dependence
on fossil fuels, it must inevitably utilise the Fast Breeder
Reactor (FBR). The reason is simple. Uranium is a finite
fuel, in just the same way as oil, coal and gas are in
limited supply, and reserves are only expected to last
sixty to ninety years at the current rates of use. As early as
1952, the physicist Samuel Glasstone observed, 'the general usefulness of nuclear fission energy, apart from
special cases, will depend to a great extent on the possibility of breeding .   . Until success in this connection is
achieved, the future of nuclear energy is somewhat in
doubt.'

Global warming threatens the world with ecological
insecurity (see chapter 4). It is important that any

Advanced nuclear reactors

Accidents at Three Mile Island (1979) and Chernobyl (1986) have generated interest in 'safer' nuclear reactors. These are intended as 'fail-safe' reactors, which won't explode like Chernobyl, or release radioactivity into the environment. Several Bills have been introduced into the US Congress and Senate to encourage research into new designs. The new designs tend to be smaller in capacity – 300 to 500MW as against 1,200MW for current PWRs – and hence more expensive. They are cooled with helium, sodium or water, and are sometimes totally surrounded by an extra pond of water just in case radioactivity leaks out. The uranium fuel is produced in less concentrated pellets, which reduces overheating problems, but makes the reactors less efficient electricity producers.

Although the US Nuclear Regulatory Commission is not prepared to accept these new approaches to safety 'as being completely adequate' at present, licenses may be secured in the next five to ten years. It will probably take a minimum of twenty-five years to have fully commercial versions in operation. The reactors will still produce radioactive waste.

solutions don't replace this with other political and environmental *problems*. Development of the Fast Breeder Reactor, as the Royal Commission on Environmental Pollution noted in its report to the UK government in 1976, poses a number of serious questions, particularly those relating to the development of nuclear weapons:

The dangers of the creation of plutonium in large quantities in conditions of increasing world unrest are genuine and serious. We should not rely for energy supply on a process that produces such a hazardous substance as plutonium unless there is no reasonable alternative.

A large nuclear programme, under any of the expansion rates calculated by Keepin and Kats, would produce a substantial amount of plutonium. While most of this plutonium would be relatively poor 'weapons-grade' material, it could still be used to make bombs. Good 'weapons-grade' plutonium comes from uranium fuel that has spent only a few months inside the reactor core, rather than the three to four years that is normal when producing electricity. The longer the fuel stays in the reactor core, the greater the proportion of plutonium-240 – one of the isotopes produced in a reactor core – which has a tendency to spontaneously fission and set off a premature chain reaction. This may lead to a less powerful but 'dirty' explosion. Dr Albert Wohlstetter, like Alvin Weinberg, a Manhattan Bomb Project scientist, confirmed in giving evidence to the Windscale Inquiry in 1977 that the US had successfully tested a weapon with 'reactor-grade' plutonium in the early 1970s.

Can we be confident that even minute quantities of the world's growing plutonium stockpile would not be diverted for purposes other than the generation of electricity, either by terrorists or governments anxious to join the Nuclear Club? History does not give us such confidence.

'The Buddha is smiling.' With this simple message, Indian Prime Minister Indira Gandhi was informed on 18 May 1974 that her country had become the sixth nuclear weapons state. The bomb had been manufactured using civil technology from Canada and other countries. In spite of vehement claims that the explosion in the Rajasthan desert near Pokharan had been 'peaceful', it immediately renewed fears that an expansion in civil nuclear power would allow the further proliferation of nuclear weapons.

South Africa and Israel are undoubtedly nuclear

weapon states. The CIA concluded, in an internal report released under the US Freedom of Information Act, that South Africa exploded a small nuclear weapon in the Indian Ocean in 1979. A former employee of the Israeli Nuclear Research Centre, who revealed details of his country's nuclear weapons arsenal to *The Sunday Times* in 1987 (estimated to be at least 200 nuclear weapons), is currently languishing in jail after being abducted by Israeli secret service agents in Italy. Other countries which have made no secret of their wish to obtain nuclear weapons include Pakistan, Argentina, Brazil, Libya and Iraq – whose Osiraq reactor outside Baghdad was bombed and substantially destroyed by Israeli jets in June 1981, because of fears over nuclear weapons development.

A small and under-resourced group of inspectors employed by the IAEA attempts to 'safeguard' the world from nuclear weapons proliferation. These 150 inspectors, aided by remote control cameras and seals on fissile materials, have to cover more than 800 facilities worldwide. The IAEA safeguards system has its limitations. It includes inspection of only those facilities to which the IAEA is allowed access. Many countries don't allow this, particularly those who are not signatories to the Nuclear Non-Proliferation treaty (NPT). The NPT was negotiated in 1965 to stop proliferation. Non-signatory countries include India, Pakistan, China and Argentina. The sanctions available to the IAEA when infringements to the NPT occur are extremely limited, and depend heavily upon diplomatic pressure from other countries. In truth, says physicist and author Walt Patterson, 'no pious allusions to "safeguards" can ultimately disguise the fact that "civil" activities involving separated plutonium are not now or may never be adequately safeguarded.' Determined terrorists, or countries willing to invest the resources and time to acquire nuclear weapons can do so.

So the nuclear genie is out of the bottle and it will be almost impossible to put it back in again. Faced with this

Uranium and the Fast Breeder Reactor

The goal of a 'plutonium economy' based on Fast Breeder Reactors (FBRs) and reprocessing plants like Sellafield has always been a kind of Holy Grail for the nuclear industry. The reason is simple. Uranium fuel is finite, and the current designs of reactors only use a small proportion of its potential energy.

According to the uranium *Red Book* produced by the OECD's Nuclear Energy Agency, 'reasonably assured resources' and 'estimated additional resources' of uranium total some 5,200,000 tonnes. A further 8,300,000 tonnes is regarded as more speculative, with 'less reliance placed on the estimates', and the price of uranium would have to increase by more than 600 per cent to make it economically viable. Over 50 per cent of the world's uranium reserves are in five countries – Australia, the USA, South Africa, Canada and Niger. Current rates of consumption are 40,000 tonnes per year, with rates of 52,000 tonnes per year expected by the year 2000. The reserves would thus last between sixty and a hundred years.

FBRs can theoretically get sixty times as much energy out of the same amount of uranium, though this has yet to be technically proven. Large sums have been spent on FBR research – $3.5 billion in Britain since 1950, and $1.6 billion in the US for the Clinch River plant, which has never operated. Breeder research has been virtually cancelled in both countries. Only one near-commercial plant has been built, the Super-Phenix near Lyons in France, and it is producing electricity at twice the cost of French PWRs.

The 'plutonium economy' would generate large quantities of potential weapon-making material. At the end of 1988, the world had about 700 tonnes of plutonium within spent nuclear fuel, 90 tonnes of which had been separated out in reprocessing plants. If nuclear capacity was to be expanded six-fold in an attempt to reduce carbon dioxide emissions, by the year 2025 the world would be producing enough plutonium for 30,000 nuclear weapons every year.

reality, should we dramatically expand nuclear power, and in particular the FBR, or should we try to control it further? Jacques Cousteau, the well-known ecologist and broadcaster, is in no doubt:

> Human society is too diverse, national passion too strong, human aggressiveness too deep-seated for the peaceful and war-like atom to stay divorced for long. We cannot embrace one while abhorring the other; we must learn, if we want to live at all, to live without both.

A risky option

The issues raised by an increased use of nuclear power are profound. In attempting to solve the problem of global warming, we may create additional safety and political problems for the world. One nuclear critic likens it to replacing a 'migraine with a stomach ulcer'.

A number of writers have asked whether the global-warming issue might reverse public and environmentalists' opposition to nuclear energy. We have argued that nuclear power can at best make only a limited contribution to reducing greenhouse-gas emissions, and even then, only if a massive world-wide energy efficiency programme holds down global energy consumption first. Without this prerequisite, nuclear power, even on a scale hitherto undreamt of by pro-nuclear advocates, would simply be an extremely expensive route for generating electricity. It would also use up potential investment capital for other policy options such as energy efficiency and renewable energy sources. It should not therefore be the highest priority for investment.

Nuclear power – indeed any other source of electricity – is not a risk-free option. Despite the confidence of nuclear agencies such as the IAEA, and the undoubted dedication and excellence of nuclear engineers and scientists, accidents will continue to occur in the future, and problems in finding *safe* disposal options for high-level waste will remain. Basing a substantial proportion of our global energy system on a fuel capable of being used in nuclear weapons is inherently and instinctively a risky option. Few people believe that any methods of safe-guarding fissile material could stop those determined to acquire nuclear weapons from doing so. IAEA Director General Hans Blix regularly and passionately argues for the dismantlement of the world's 50,000 nuclear weapons, in order to remove the 'Sword of Damocles' from above our heads. The development of the FBR and the spread of plutonium is likely to undermine this very process. At a time when the first faltering steps towards world nuclear disarmament are being made, and major thaws are occurring between ideologically divergent blocs, less risky and politically decisive alternatives should have a higher priority.

Over the past thirty-five years, nuclear power has been given unparalleled financial and political support from governments throughout the world (see 'A short history of nuclear power', pp. 174–175). The fact that it still provides less than 3 per cent of the world's delivered energy, often at a high cost, should give us cause to reflect, especially as there remain so many problems that need to be solved. We should question the motives of those who advocate a rapid expansion of nuclear power as a solution to global warming. Is it a genuine expression of concern? Or is it simply a desperate attempt to salvage nuclear's fading star on the back of the world's most serious environmental problem?

7

SAVING THE PLANET

My experience of the world is that things left to themselves don't get right.

Thomas Henry Huxley

This chapter is unashamedly about action – action to save our planet from bigger climatic changes than have been seen for 100,000 years. The authors have not hesitated to point out that there are uncertainties. We are convinced, however, by both the evidence available to us and the strong support by scientists, of the need for action. We do not claim to know all the answers. Nor can we suggest that all of the proposals will have a successful outcome. We simply take our lead from those like the United Nations Environment Programme (UNEP) Deputy Director, Genady Golubev, who maintains that: 'advocating patience is an invitation to be a spectator at our own destruction'.

Strong statements from the Toronto Conference and the Hamburg Congress have already been noted in chapter 5. A call for action has come from well-informed politicians such as Norwegian Prime Minister Gro Harlem Brundtland, Brian Mulroney, Prime Minister of Canada, and Helmut Schmidt, the former West German Chancellor. Major scientific organisations such as United Nations Environmental Programme (UNEP), the World Meteorological Organisation (WMO) and the International Council for Scientific Unions (ICSU) are calling for action now – while further research and monitoring is carried out – and have initiated high-level policy discussions on the steps to take. A number of countries such as the Netherlands and Norway have unilaterally committed themselves to holding down energy consumption and carbon dioxide emissions.

This chapter is about the co-ordinated strategy and specific actions needed to save our planet.

The four track strategy

The World Commission on Environment and Development, in its report *Our Common Future* (see Brundtland Commission – chapter 4), concluded that in view of 'the complexities and uncertainties surrounding the issue (of global warming), it is urgent that the process of action starts now'. They recommended a four-track strategy which combined:

☐ improved **monitoring** and assessment of the evolving climatic changes and systems;
☐ increased **research** to improve knowledge about the origins, mechanisms, and effects of these changes and systems;
☐ the development of internationally agreed policies for the **reduction** of the greenhouse gases;
☐ **adoption** of strategies needed to minimize damage and cope with the climate changes and rising sea level.

This framework will form the basis of 'Saving the Planet'.

Although good progress is now being made towards the first two goals, more needs to be done. Little has so far been proposed or carried out on the final two goals however, which are essential for the maintenance of life as we know it.

The Brundtland Commission identified the many conflicts and inequities which are the root cause of increasing

The Four Track Strategy

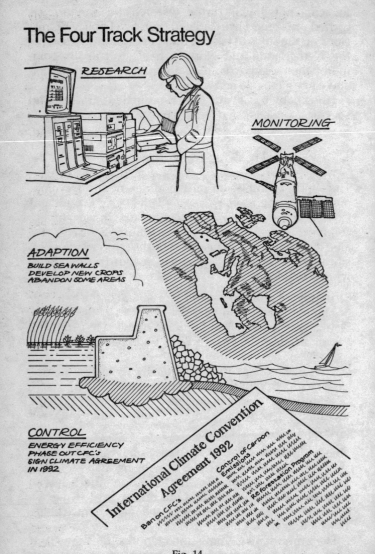

Fig. 14

environmental degradation and poverty in so many
countries; one of the themes in chapter 4. They further
developed the concept of 'sustainable development' as a
goal for all nations, and argued that this lies in the cooperative management of resources among nations,
good communications to lessen uncertainty and insecurity, and a redefinition of priorities, nationally and
globally, away from arms and other wasteful uses of
resources, and towards the environmental tasks ahead.
As they point out:

> Four of the most urgent global environmental require-
> ments – relating to tropical forests, water, desertifica-
> tion and population – could be funded with the
> equivalent of less than one month's global military
> spending.

The world spent an estimated $3 billion a day on military
purposes in 1988.

Chapter 5 showed that there are a range of options
available to countries for slowing global warming. Any
actions proposed should be based on a number of general
principles. These include:

☐ **resilience** – the strategy should be able to achieve
results even under uncertain and changing circumstances;

☐ **avoidance of irreversibility** – any actions should
avoid establishing an irreversible global system;

☐ **a global outlook** – an international perspective is
needed that puts the common interest above national,
individual or regional considerations. Some countries
may have to forgo advantages to allow others to avoid
environmental and economic damage;

☐ **equitability** – the actions should always try to dis-
tribute the costs and benefits equitably – between and
within countries – with a bias in favour of the less
developed countries.

The actions also need to take account of our current knowledge of the greenhouse gases, the extent to which we have technology available to reduce the emissions, and the likely impact of any reduction strategies. Such an assessment identifies CFCs as our number one priority.

According to Chris Rose, an experienced environmental campaigner who advises Greenpeace International:

> CFC sources are readily known and understood as is the fate of the gases in the atmosphere. Substitutes and the technology to recycle the gases are generally available, and the economic penalties of changing are quite small. The consequences of delay are so serious [see chapters 2 and 3] that justification for rapid policy action is extremely high.

It has been estimated that alternative gases for refrigerators might put an extra 50p on the cost of each unit.

Carbon dioxide emissions are another high priority. We already have a good overall understanding of the carbon cycle, its sources and its sinks, but major uncertainties remain about the possible take-up of carbon dioxide by the oceans and the flux to and from the soils. The technologies for reducing carbon dioxide are known, but options such as building new power-stations can take a long time to implement. We therefore have to start planning them now.

We know much less about the sources, cycles and sinks of nitrous oxide and methane. Technology to reduce emissions is only partially known or available, though reductions in fossil-fuel consumption would help. Tropospheric ozone can be reduced by fitting catalytic convertors to car exhaust pipes. Since ozone only lasts a few weeks, the results would be almost instantaneous.

Monitoring and research

Monitoring is essential if the first effects of global warming are to be detected accurately and in time. Without hard and accurate information, the most sophisticated computer model will produce inaccurate results. Early warning is needed so we can take action.

There is a lack of clear co-ordination of monitoring information and systems. The UNEP Global Environmental Monitoring System (GEMS) is a modest initiative. GEMS, and its associated Global Resource Information Database (GRID) and 'Earthwatch' programmes, should be significantly expanded. Other programmes need more resources to extend the number of monitoring sites and the amount of information they provide. These include the Geosphere-Biosphere Programme run by the International Council of Scientific Unions (ICSU); and the United Nations Disaster Relief Office, which gives early warning of imminent natural disasters.

Action check list for governments and international agencies

☐ Extend the monitoring of greenhouse-gas emissions (particularly of methane and nitrous oxide), the levels and behaviour of carbon dioxide in the oceans, and sea-levels.
☐ Extend the recording of rainfall and moisture throughout the world.
☐ Increase and extend the level of research into cloud behaviour, the roles of vegetation and ice cover, and the relationship between albedo, weather and climate.

A clearer understanding of climatic feedbacks should be the overall objective of any extended research and monitoring programme.

Controlling emissions

New evidence from the US Environmental Protection Agency (EPA) has confirmed that a near 100 per cent reduction in fully-halogenated CFCs, plus a freeze on methyl chloroform, is needed to stabilize stratospheric ozone. The UK Ozone Trends Review Panel, chaired by John Pyle, has accepted this recommendation, and it is now official government policy.

Robert Watson of NASA told a scientific conference organised by Friends of the Earth in November 1988 that the Antarctic ozone hole would not disappear until atmospheric chlorine from decaying CFCs was reduced to around two parts per trillion by volume (pptv). This, he said, would require 'an almost complete phase-out of CFCs and careful consideration of other chlorine carriers'. Only if participation in a re-negotiated Montreal Protocol improved to near 100 per cent would chlorine levels be stabilized by the year 2100.

There is now a clear unanimity that the Montreal Protocol should be reassessed – and fast. Several years ago, major CFC producers such as Du Pont and ICI argued for delays in banning the substances due to the supposedly long lead times in developing alternative chemicals. This is no longer the case. Non-chlorine carrying compounds are either already available, or are expected to be ready soon, for uses as solvents, aerosols, metal cleaning and foam-blowing agents. 'Soft' CFCs (CFCs only 5 to 10% as harmful as CFC11 and 12) are already being used as foaming agents. Hydrochlorofluorocarbons such as HCFC123 and HCFC141b, which break down in the lower atmosphere, are already being tested for toxicity as substitutes for CFC11 in making rigid foams. Alternative foam-blowing agents include methylene chloride, pentane and carbon dioxide. Each have their drawbacks: the first is carcinogenic, pentane is flammable, and carbon dioxide is a poorer insulator.

Other technologies exist for refrigeration and air-conditioning, though these will probably prove to be the industries most resistant to substitutes. Chlorine-free HFC134a is the main substitute available at present, though this is a powerful greenhouse gas. Helium, though more expensive, is another promising substitute for small refrigeration units, and ammonia and lithium bromide can be used for larger units in supermarkets and elsewhere.

There is a very large halon 'bank' in fire-estinguishers – 70 to 80 per cent of all the halons ever produced. UNEP points out that only 4 per cent of Halon 1211 and 7 per cent of 1301 are used on fires: 12 and 23 per cent respectively are lost in testing fire-extinguishers. At least half of these losses could be stopped almost overnight by changes in testing procedures.

At the end of an international technical conference on CFC substitutes in October 1988, UNEP concluded that there were 'no valid technical reasons' for not moving from CFCs.

Achieving a near complete reduction of CFCs will require a coordinated strategy. Alan Miller and Irving Mintzer of the World Resources Institute in Washington suggest a four-fold programme. The first goal is to reduce leakages by increasing efficiency in CFC production and use. In car air-conditioning, about 30 per cent of CFC12 used is lost by leakage and escapes during servicing. Better designed equipment and more careful servicing would reduce leakage significantly.

The second goal of the strategy is to encourage CFC recovery and recycling. A two- to three-fold price increase is needed to encourage recycling in small de-centralised uses such as vehicle air conditioners. In factories, the economics of recycling are already good.

The third goal is to use 'soft' CFCs, those which have a much lower ozone depleting and greenhouse-warming potential. Substitutes include HCFC22, HCFC124, HCFC141b, HFC134a and HFC152a. Du Pont has already produced blends of these for the domestic fridge/freezer and air-conditioning market. For example CFC134a does not harm the ozone layer, and has only 10 per cent of the global-warming potential of CFC11 or CFC12 (see appendix 5). However, if production and consumption were to increase rapidly in the developing world, we would be back to square one. In Beijing, China, for example, the percentage of households owning a fridge has increased from 3 per cent to more than 60 per cent in the last five years. 'Soft' CFCs should be regarded as an interim measure only with a clear phase-out date of 1995 at the latest. Vigorous diplomatic efforts should be made to dissuade countries such as China and India – who argue that they should not be held responsible for destroying the ozone layer – from using CFCs regulated by the Montreal Protocol. The transfer of CFC-substitute technology to these countries will be crucial components in their willingness to sign a reassessed Montreal Protocol.

The fourth goal is to switch to processes and products which require no CFCs at all. Hydrocarbon propellants such as butane are available for most aerosols and already meet 90 per cent of the US market. UK aerosol manufacturers are committed to achieving similar levels by the end of 1989. Substitutes for CFC-blown insulation materials are available, including fibreglass and expanded cellulose, cardboard packaging competes well with polystyrene foams. Methyl chloroform or methylene chloride can replace CFC113 as a solvent.

Check list for consumers

☐ Use the list of CFC-free aerosols published by Friends of the Earth groups. FOE UK can provide information on aerosols, foam packaging, building insulation and other products.

☐ Avoid closed-cell foam packaging for eggs, burgers and other food.

☐ Check that any furniture you buy does not contain open-cell foams with CFCs. High-density foams are a little harder and more expensive, but are either CFC-free or use only small amounts.

☐ Write to the fridge and freezer manufacturers demanding that they use CFC substitutes such as HCFC22. Aim to buy the most efficient fridge or freezer – consult the Consumers Association's *Which?* magazine for details of the electricity consumption.

☐ Avoid air-conditioning in cars. This is not necessary in temperate climates such as Britain's. If you have air-conditioning, get any leaks seen to immediately. Don't replenish the system without repairing the leak.

Priorities for government and industry

☐ Re-negotiate the Montreal Protocol to enforce a phase-out of fully-halogenated CFCs and other ozone-depleting substances by 1995 at the latest.

☐ Clearly label all products using CFCs.

☐ Develop mandatory CFC recycling and recovery schemes for factories, garages and industrial refrigeration systems, and seek out substitutes.

☐ Major CFC manufacturers and host countries should negotiate 'technology-transfer' schemes with developing world countries for CFC substitutes which are less harmful to the ozone layer and global warming, as an interim measure until 1995. Britain's Overseas Development Agency (ODA) has an important role in encouraging this, as well as ensuring that it does not fund projects adversely affecting the climate.

☐ Car manufacturers should offer vehicles without air conditioning, particularly in North America and Japan.

☐ Companies should avoid test discharges of halon fire-extinguishers. 'Total flooding' tests in places like computer rooms are intended to identify leaks. The same thing can be achieved with a large fan.

Carbon dioxide – five billion producers

Getting rid of CFCs is a relatively small and simple problem. Reducing carbon dioxide emissions is much more complex. Energy consumption and land use changes are crucial components of economic development, and solutions to our excess production of carbon dioxide involve fundamental issues of equity and sustainable development.

Planting trees for survival

> He who plants trees loves others beside himself.
>
> English proverb.

Several scientists have advocated planting large numbers of trees to 'soak' up excess carbon dioxide. The work by Greg Marland, from Oak Ridge National Laboratory in the US, showed that a forest of 500 million hectares – the size of the US (minus Alaska) – would be required to soak up half of the carbon produced by burning fossil fuels each year. This is a lot of trees – or is it? Although it would require the planting of 18.4 billion trees annually up until the turn of the century, this amounts to every one of us planting and caring for just four tree seedlings a year. As Worldwatch Institute researchers Sandra Postel and Lori Heise state, 'reforesting the Earth is possible, given a human touch'.

In October 1988, a small American electricity company, Applied Energy Systems, from Arlington, Virginia, announced that it had arranged with aid agencies and the government of Guatemala to plant 15 million trees in an area of about a thousand square kilometres. The company had calculated this was the amount of forest needed to absorb the 387,000 tonnes of carbon dioxide per year from their new 180MW coal burning power-station.

It was, World Bank ecologist Robert Goodland told a London conference the following month, the first time anyone had 'internalised voluntarily . . . the world's most persuasive negative externality (carbon dioxide) . . . a wonderful guide for the future.' He added:

> This suggests that greenhouse-gas emissions, voluntarily or by taxation, could be used to subsidize the sink function of the environment for their absorption. Carbon sink forests cannot solve the greenhouse problem, but they can certainly buy time for its solution . . . thirty to forty years of time to reduce greenhouse gases.

The links between industrialised and developing countries are also clear from this example. The developed world contributes more to the problem of climate change, but as we have seen in chapters 3 and 4, the developing world will suffer more. Atmospheric carbon releases could at least subsidise tropical afforestation to global benefit.

The provision of an additional carbon sink through rapid re-afforestation does have a role to play in slowing global warming, as part of a set of policy actions, but tree planting must take account of local ecology. Some areas, such as the Flow Country in northern Scotland, are not really suitable. Individually, we should all do what we can to plant trees every year, either through the many Nature Trusts and organisations such as the Tree Council and 'Men of the Trees'.

Unless the destruction of Amazonia and other tropical forests is halted, tree planting will have a limited impact. How is this to be done? Chapter 5 explored some of the possibilities. We propose the following actions to preserve the tropical forests and encourage re-afforestation schemes.

Managing the forests

How are the rainforests to be saved? Leading conservation organisations like the World Wide Fund for
Nature are now convinced that it cannot be done by
halting their exploitation, particularly not the tropical
hardwoods trade which is driving some of the forest
destruction in South-East Asia and Africa. Brian
Johnson, an independent forestry consultant who advises WWF and other organisations, says 'if the trade
is abandoned, the producer countries will abandon
the forests to agriculture.' Logging is directly responsible for only a small proportion of forest loss, but the
indirect effect is greater, since the operation opens up
the forest to settlers.

Duncan Poore, a former director of the Commonwealth Forestry Institute in Oxford and an adviser to
the International Institute for Environment and Development, says:

> It is inconceivable that nations could justify making
> over 800 million hectares of forest into protected
> forests: if it is not managed for economic purposes it
> will certainly disappear as soon as it becomes access-
> ible.

In 1987, the FAO, the World Bank, the UN Development Programme, the World Resources Institute and
the Rockefeller Foundation set up a five-year Tropical
Forestry Action Plan (TFAP) aimed at promoting the
conservation and sustainable use of the remaining
forests. The plan's promoters point to many successful
enterprises by aid agencies, mobilising local people to
preserve and restore forests and wooded land. But the
task is immense. Only 40 per cent of the world's
natural forest that is potentially productive is being
managed well enough to sustain even its current yield.
No more than a million hectares is sustainably managed for logging, less than 1 per cent of the total. Only

Trinidad and Tobago, and Malaysia, whose practice does not always match its principles, adopt a management approach at all.

The TFAP has its critics. Vandana Shiva, of the Indian Chipko ('tree huggers') movement, regards it as colonialist and a 'top-down' approach in which foreign consultants and development banks tell people what is good for them. Koy Thompson, Friends of the Earth's International Tropical Forest campaigner, agrees. Thompson argues that the Plan's solution of increased investment and expenditure on forestry, often leads to the removal of tribal people from their land and the rapid demise of the forest.

Time is short for the world's rainforests. Despite its failings, the TFAP at least provides a basic structure for discussion and the setting up of appropriate economic measures to ensure sustainable forestry.

Action check list for the rainforests

☐ Individual countries and international aid agencies should assess the potential of tree-planting schemes for carbon sinks; overseas aid agencies should include the role of forests in climatic change as a criteria when assessing investment projects.

☐ Policies should be developed for encouraging organisations and individuals to plant trees, subject to assessing their ecological and social impacts.

☐ Industries producing large quantities of carbon dioxide, or those who clear large areas of forests, should be obliged to include the costs of this pollution in their investment calculations, or at least compensate by planting 'carbon sink' forests.

☐ 'Debt-swap' arrangements, whereby debt is written off for protecting areas of rainforest, should be en-

couraged by aid agencies where appropriate, in close
consultation with both governments and NGOs.

☐ An international regulation should be introduced
which would require that consumer countries only
buy hardwoods from sustainably managed forests,
and producer countries adhere to a code of conduct on
'best practice' for forest management.

☐ Builders and other major consumers of tropical hardwoods should be required by law to buy from sustainably managed forests.

☐ Compulsory labelling of hardwoods to indicate the
country of origin and whether it has come from a
sustainably managed forest, should be introduced.

☐ Individual consumers should buy only those tropical
hardwoods from sustainable forests with a 'Good
Wood' seal of approval. (Details from Friends of the
Earth.)

The energy problem

Up to 80 per cent of man-made atmospheric carbon
dioxide comes from fossil fuel. Oil, coal and natural gas
are the driving forces of the industrial age, consumed in
countless billions of cars, furnaces, boilers, chemical
plants and cookers throughout the world. Six billion
tonnes of carbon are released each year producing nearly
20 billion tonnes of carbon dioxide. The world needs to
cut these emissions by at least 50 per cent in order to keep
the rate of global warming to 0.1°C/decade or less.

The warming commitment model – implications for energy policy

Chapter 2 indicated the possible range of global temperature increases by the middle of the next century – from 0.06°C per decade to a massive 0.8°C per decade, if greenhouse-gas emissions carry on at present rates. Temperature increases at anything other than the lower end of the scale are likely to lead to significant disruption of ecosystems and human settlements, according to the 1987 Villach–Bellagio conferences (see chapter 2 and Appendix 1). A goal of 0.1°C per decade was therefore selected as the warming rate which would allow natural ecosystems and societies to adapt, and which had previously been experienced in history.

The extent to which current energy and industrial policies must change has been demonstrated by Irving Mintzer's **Warming Commitment Model**, which simulates the effects of various policies on the build-up of greenhouse gases in the atmosphere. Mintzer, from the World Resources Institute, developed four possible scenarios for the model. These included a 'High Emissions' scenario, a 'Base Case' scenario (which adopted a 'business-as-usual' approach), a 'Modest Policies' scenario which reduced emissions to some extent and a 'Slow Build-up'. The results are stark. Only in the 'Slow Build-up' scenario is a doubling of carbon dioxide levels prevented over the next ninety years. In all other scenarios, future warming commitments in relation to pre-industrial levels ranged from 1.8 to 10°C by the middle of the next century. This equates to a warming rate of 0.1 to 2.0°C per decade. Even under the 'Slow Build-up' option, the Earth's surface temperature is committed to an increase of 1.3 to 3.8°C by the year 2050, a warming rate of 0.1 to 0.3°C per decade.

To achieve a slow build-up, Mintzer argues that the pricing of energy must reflect its high environmental costs, and hence encourage a switch away from solid fuels to natural gas. Energy efficiency improvements at a rate of at least 1.5 per cent per year, and increased research support, tax incentives and other subsidies for solar energy are also needed. There are several ways of doing this. None is easy. Chapter 5 has shown that they depend on how much energy people will need in future.

The high-energy scenarios discussed in chapter 5 cannot be regarded as appropriate for a strategy seriously intent on reducing greenhouse-gas emissions. There is no historical imperative that energy growth must continue indefinitely into the future. 'There is no inherent worth in a lump of coal, a barrel of oil, or a tank-full of gasoline,' says Amory Lovins, Director of the Rocky Mountain Institute in Colorado. 'What people really need is the energy service these fuels can provide in the form of heat, light or motor power.' For the past sixteen years, Lovins has argued, written and lectured extensively throughout the world on the benefits of the 'soft energy path' which aims to provide these services as efficiently as possible.

BP's managing director, Robert Malpas, soon to become chairman of one of the UK's privatised electricity utilities, agrees: 'I am struck by the convincing and attractive assessments of what greater efficiency can achieve, and how little it figures in major debates and policy on energy.' Malpas points out that part of the problem is that there is no way at present for the long-term environmental consequences to be reflected in energy investment decisions.

If there were some way of bringing home this cost to a present value for the public, they might opt for greater environmental protection today. A *Daily Telegraph* opinion poll in 1988, which showed a significant increase over six years in favour of protecting the environment rather than keeping prices down, suggests that the public are indeed prepared to pay for this.

Energy in the UK

KEY FACTS
- ☐ The UK energy bill is £39 billion, 9 per cent of Gross Domestic Product (GDP).
- ☐ The government estimates that at least 20 per cent of this is wasted – equivalent to 150,000 new houses, or 22 major hospitals each year. Government scientists estimate that even bigger savings of up to 40 per cent are technically possible.
- ☐ Most of the measures to save this 20 per cent are straightforward, highly economic, and involve no loss of comfort.
- ☐ 65 per cent of our energy spending at home goes on space and water heating, 9 per cent on lighting, 12 per cent on electrical appliances, and 6 per cent on cooking.

ENERGY SAVING ACTION PLAN
- ☐ Between 15 and 50 per cent of heat is lost through draughty windows, doors and floorboards. Use brushstrips around letter-boxes and at the foot of doors, rubber and plastic seals around doors and windows, and a sealant for gaps between floor and skirting boards. At a cost of £30 to £80 all in, feel the extra warmth and savings will be felt immediately (remember to turn down your thermostat though).
- ☐ Insulating your loft and hot water-tank is also cheap – £80 to £150 depending on the size of the house. If you already have some loft insulation, top this up to six inches.
- ☐ Solid-walled homes can be insulated on the outside or inside, using foam or polystyrene material (make sure it doesn't contain CFCs).
- ☐ External or internal shutters, especially when insulated, are very good heat conservers.
- ☐ Check your radiators to see whether they have thermo-static valves, allowing you to adjust the temperature in each room. Get them fitted by a plumber at your next heating service.

☐ New gas condensing boilers are 25 per cent more
efficient than conventional ones – check with your
local gas showroom.

☐ Double glazing isn't really cost-effective in energy
saving terms, but if you've insulated the house
well, fitting 'secondary glazing' or replacing old
windows with double glazing makes a lot of sense.

☐ Remember – you can get lots of help and advice
from the Energy Efficiency Office and Neighbour-
hood Energy Action (see appendix 6) and your local
electricity or gas showrooms which are supposed
by law to provide assistance.

Using fossil fuels more efficiently

Given that many developed countries depend on fossil
fuels for 90 per cent of their commercial energy, it will be
impossible to switch overnight from fossil fuels to
carbon-free energy sources, whether these be nuclear
power or renewable energy sources. Improvements in
the efficiency of using fossil fuels is a high priority and
there is much we can do to use them in a cleaner and
more efficient way. Power-stations, which produce 30
per cent of global carbon dioxide emissions and 38 per
cent in Britain, are often only 35 per cent efficient. 'Over
20 per cent of the UK's carbon dioxide emissions come
from power-station waste heat which provides no benefit
to the public or industry whatsoever,' says David Green,
Director of the Combined Heat and Power Association.
Combined heat and power (CHP) stations double the
efficiency of power-stations by piping the waste heat to
homes, offices and industry. Around 10 per cent of
Britain's total energy demand could come from this
'waste heat', saving up to 50 million tonnes of carbon

dioxide. Green points to CHP schemes in major cities such as Berlin, Milan, Paris, Rotterdam, Moscow, New York and Copenhagen as a sign that CHP is a proven technology.

The International Energy Agency lists at least thirty technologies which can help clean up coal and use it more efficiently. Technologies such as 'Fluidised-bed combustion' and 'Integrated Gas Combined Cycle' are now available in a range of power-station sizes, with efficiencies of up to 55 per cent. All have much lower acid emissions than conventional stations. Walt Patterson, author of a recent report on advanced coal-use technology, puts the case for cleaner coal succinctly when he says that 'the immediate objective must be to get as much useful energy as possible from every kilogramme of fossil fuel we turn into carbon dioxide'. The same must be true of the fuel consumed by the car.

Rethinking the car

Cars and light vans produce around 18 per cent of global carbon dioxide emissions; they are also major contributors to tropospheric ozone. The car is nothing short of an environmental disaster (see 'Some Key Facts about the Car', p. 210). It is also the fastest growing source of carbon dioxide. In Britain, emissions are growing at 4 per cent per year, and action to control them is an urgent priority. Measures to do this include planning communities in which people can live closer to their workplace, providing cheap and extensive public transport, encouraging the use of bicycles, and heavily taxing personal vehicles. This sounds like a curtailment of essential freedoms, but such measures are unavoidable due to the rapidly worsening environmental impacts in cities and elsewhere. Restrictions on car use are already in operation in Athens, Milan and a number of other cities to

reduce congestion and air pollution, which in the words
of Kurt Patrick Fale, act like an 'urban thrombosis that
slowly deprives the city of its life-blood.'

Some key facts on the car

- [ ] There are 400 million cars and light trucks world-
  wide, compared to 50 million in 1950. More than 30
  million are built every year.
- [ ] 16 per cent of the world's population in North
  America, Europe, Japan and Oceania produce 88
  per cent, and own 81 per cent, of all cars.
- [ ] The US produces 25 per cent and owns 35 per cent
  of the world's cars.
- [ ] 18 per cent of global carbon dioxide emissions are
  from cars.
- [ ] The car causes 30,000 deaths in the US and over
  5,000 in the UK each year. Air pollution results in
  additional traffic congestion and delays, and health
  effects due to lung and heart damage.
- [ ] Ozone pollution causes up to $4.5 billion damage to
  US crops each year.
- [ ] 50 per cent of vehicle trips in the US only involve
  the driver.
- [ ] Average speeds are 10mph in London and even
  less in Tokyo. Average speeds in Southern California, the most car-dominated society in the world,
  are expected to fall to 15mph by the year 2000.
- [ ] In the developing world, only 1 per cent of the
  population own a car, compared to 40 per cent in
  the industrialised West.
- [ ] The global average efficiency of vehicles is 18 miles
  per gallon (mpg), and 36mpg in Japan and Western
  Europe.
- [ ] Weight reduction, new engine designs, variable
  transmissions and other improvements could
  easily double fuel efficiency by the year 2000.

Jeremy Vanke of Friends of the Earth argues that the huge subsidies provided for the car – $300 billion each year in the US and £2.5 billion ($4.5 billion) for company cars alone in the UK – place public transport at a severe disadvantage. 'Getting control of the runaway car means cutting these subsidies,' he says, 'discouraging its use in urban areas for short trips, and developing an integrated public transport system.'

Doubling the fuel efficiency of cars and light trucks over a decade is technically possible, according to Deborah Bleviss of the International Institute for Energy Conservation in Washington. To secure this improvement, governments should increase fuel taxes and fuel economy standards. Most major car producers have prototype models which give fuel efficiencies in the range of 80 to 100 miles per gallon. This is achieved through measures such as weight reduction, engine and transmission efficiency improvements, reduced aerodynamic drag and tyre-rolling resistance, and electronic 'stop-start' systems which cut petrol wastage when the car is idling. Low fuel prices will not encourage the early introduction of these prototypes however.

Planning is the key for transport systems. Much of the USA is scarred for the want of transport and urban planning. Sprawling communities, gross congestion in city centres and motorways, and 80 million Americans living in areas exceeding ozone-pollution standards all testify to that. Despite this harsh reality of a car-dominated society many developing countries are starting to embrace the car as a symbol of development and progress. It may not deliver on this promise. As Vanke points out:

The car promised us all unlimited accessibility; what we've got instead is lots of mobility, much of it spent in traffic jams, road diversions, and looking for parking spaces. We've lost sight of our original goal. The car has now just become an extremely inefficient and

polluting symbol of wealth, virility and status rather
than an efficient transport medium.

Saving energy worldwide

The basis of the *Energy For a Sustainable World* scenario
described in chapter 5, which was 'neither a projection
nor a policy prescription . . . rather an illustration of
what [was] technically possible' is energy efficiency in all
sectors, a greater emphasis on natural gas than coal, and
a modest increase in the use of renewable energy
sources. The study is packed full of examples where
efficiency gains have already occurred: from advanced
fuelwood stoves in Kenya to super-insulated houses in
northern Canada heated by a few dollars of fuel.

   In one of life's sad ironies, a poor Tanzanian farmer
scratching out a subsistence living uses nearly fifteen
times as much energy to cook as a wealthy UK home-
owner using natural gas. While new designs have doubled the efficiency of cooking stoves, in many parts of
the developing world such as Kenya, the use of small
amounts of modern liquid or gaseous fuels from biomass
could dramatically reduce the need for firewood and give
a better energy service with lower fuel inputs. Dr
Ravendra Pachauri from the TATA Energy Research
Institute in New Delhi estimates that a mere 9 million
tonnes of kerosene, if burned in stoves at 60 per cent
efficiency, could cut biomass consumption in the whole
of India by half.

The means of achieving significant energy efficiency
gains are extremely varied. They range from technological change – using more efficient appliances or industrial
motors, improved building codes to increase the thermal
insulation and orientation of buildings in order to use
passive solar energy – to major changes in energy pricing
structures and mechanisms. The 'National Appliance

Energy Ecotopia

Ten years ago Ernest Callenbach wrote two novels, *Ecotopia* and *Ecotopia Revisited*, charting the progress of the west coast of America, which broke away from the rest of the USA to live a sustainable, ecological lifestyle. The books describe in great detail the transport, work, food and waste systems where everything biodegrades and the energy inputs are extremely low. Callenbach's vision is coming sharply back into vogue and popularity in the face of major environmental problems in the USA, ranging from air quality, water pollution, and toxic dumps to disappearing wildlife.

Milton Keynes may be an unlikely setting for Ecotopia, but it is certainly the UK's 'Energy City'. More than 750 energy-efficient houses have been built over the past ten years with large south-facing windows and good insulation to take advantage of the sun's free heat. Construction costs were less than 1 per cent higher than conventional houses but heating bills are usually 40 per cent lower.

In 1986 the New Town Development Corporation devised an energy cost index (ICE) and invited fifty builders to design and build energy efficient houses for an Energy World exhibition. To qualify, the houses had to meet an ICE limit of 120 (the average for UK houses was 170). Many came in well under target: 25 were below 100, 12 below 75. The lowest was 41.9 – four times better than the average British house. Steve Fuller, Director of the Milton Keynes Energy Foundation, puts it down to 'good design, the utilisation of simple but effective insulation measures, and the challenge of innovation in a competitive market'. All the houses were sold at a premium and 500 more have been built in the Energy Park over the past two years.

> David Olivier, Director of Energy Advisory Associates in Milton Keynes, says:
>
> Using currently available technology Britain could reduce its average electricity demand by 70 per cent, from 28 gigawatts to 8 gigawatts. The cost would be about 80 per cent less than generating the same amount of electricity from new power-stations.
>
> Olivier quotes the example of a modern fluorescent light-bulb, which uses 80 per cent less electricity than a traditional incandescent bulb, and saves about a tonne of carbon dioxide in its lifetime. The best mass-produced fridges and freezers are two to three times more efficient than the average. Labelling appliances and developing regulations for improved efficiency would give consumers a real choice and significantly help reduce carbon dioxide emissions.

Efficiency Standards Act' in the US is expected to reduce electricity consumption in the year 2000 by 53 TWh and peak demand by 22,000 MW (equivalent to 22 large coal/oil power-stations). A 'carbon tax', which would be varied to favour the lower carbon dioxide emissions from natural gas as compared to coal (about 40 per cent less), has been proposed by a number of policy-makers. It is crucial that some form of 'environmental tax' be placed on vehicle fuel in the USA and Canada, where prices are only one-quarter of those in Western Europe and the incentive to save fuel or introduce efficient 'gas sipper' models is low. The introduction of such a tax would clearly face some resistance from the coal and oil industries. It is essential too that the full costs of nuclear power are shown, including insurance liability for accidents and the long-term costs of waste disposal and station decommissioning, in order that nuclear power does not gain an unwarranted economic advantage.

Action check list

- [ ] Introduce a carbon tax on fossil fuels, on a pro rata level to their carbon dioxide emissions – ie, higher for coal and lower for gas. The tax receipts should fund global warming research, the development of carbon-free energy sources, and tax incentives and other fiscal measures for energy efficiency and renewable energy.
- [ ] Introduce a vehicle fuel tax to stimulate more efficient vehicles, to discourage non-essential uses, and to remove all hidden subsidies such as 'company car' tax concessions.
- [ ] Require electricity and gas utilities to invest in energy conservation by legislation which forces them to evaluate both efficiency and supply options and choose the cheaper option. This is called 'least-cost planning' in the US where it is commonplace.
- [ ] Develop higher standards of efficiency for electrical and gas appliances through legislation and labelling on the lines of the US 'National Appliance Efficiency Standards Act'. Labelling should include both the running costs of each appliance and the pollution output.
- [ ] Introduce new standards for cars and light duty trucks which double the current fuel efficiency by the year 2000.
- [ ] Develop integrated public transport systems through greater investment, and encourage their use through fare subsidies and business taxes.
- [ ] Modify building codes and regulations to improve thermal insulation, increase heating system efficiencies, take advantage of passive solar heating, and make heating controls and thermostats mandatory.
- [ ] Governments should provide extensive information and advice on energy efficiency for businesses, local government and the general public.

Check List for Action

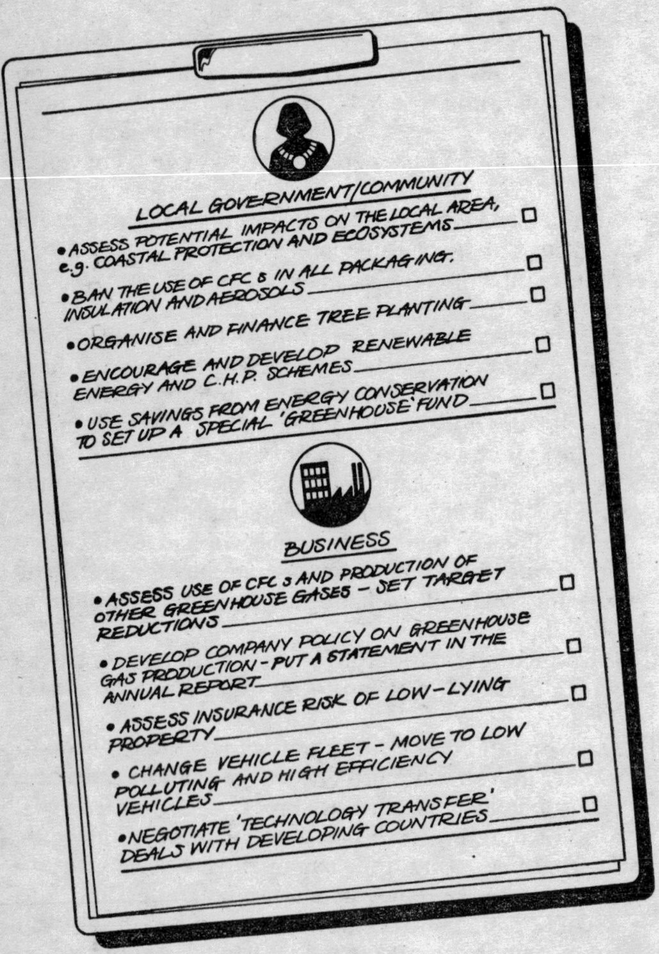

LOCAL GOVERNMENT/COMMUNITY

• ASSESS POTENTIAL IMPACTS ON THE LOCAL AREA, e.g. COASTAL PROTECTION AND ECOSYSTEMS ☐

• BAN THE USE OF CFC s IN ALL PACKAGING, INSULATION AND AEROSOLS ☐

• ORGANISE AND FINANCE TREE PLANTING ☐

• ENCOURAGE AND DEVELOP RENEWABLE ENERGY AND C.H.P. SCHEMES ☐

• USE SAVINGS FROM ENERGY CONSERVATION TO SET UP A SPECIAL 'GREENHOUSE' FUND ☐

BUSINESS

• ASSESS USE OF CFC s AND PRODUCTION OF OTHER GREENHOUSE GASES — SET TARGET REDUCTIONS ☐

• DEVELOP COMPANY POLICY ON GREENHOUSE GAS PRODUCTION - PUT A STATEMENT IN THE ANNUAL REPORT ☐

• ASSESS INSURANCE RISK OF LOW-LYING PROPERTY ☐

• CHANGE VEHICLE FLEET - MOVE TO LOW POLLUTING AND HIGH EFFICIENCY VEHICLES ☐

• NEGOTIATE 'TECHNOLOGY TRANSFER' DEALS WITH DEVELOPING COUNTRIES ☐

Fig. 15

NATIONAL GOVERNMENT

- SET TARGETS FOR CO_2 REDUCTIONS – AT LEAST 20% BY THE YEAR 2000 ☐
- PHASE OUT CFCs BY 1995 ☐
- INTRODUCE 'LEAST COST' PLANNING FOR ELECTRIC AND GAS UTILITIES ☐
- SET NEW STANDARDS FOR VEHICLES AND APPLIANCES – ABOLISH COMPANY CAR SUBSIDIES ☐
- SET NITROGEN QUOTAS IN AGRICULTURE ☐
- INTRODUCE A CARBON TAX ON FOSSIL FUELS ☐
- ASSESS ALL AID PROJECTS FOR THEIR GLOBAL WARMING IMPACT ☐
- CHANGE EMPHASIS OF ENERGY R&D BUDGET TOWARDS ENERGY EFFICIENCY AND RENEWABLES ☐

INTERNATIONAL

- RE-NEGOTIATE MONTREAL PROTOCOL TO PHASE OUT CFCs BY 1995 ☐
- AGREE AN INTERNATIONAL CLIMATE CONVENTION BY 1992 WITH PROTOCOLS TO REDUCE GREENHOUSE GASES – ESPECIALLY CO_2 ☐
- INCREASE BUDGET AND SCOPE OF MONITORING AND RESEARCH ☐

There is a fairy-tale about the sun and the wind looking down at the Earth and seeing a freezing man wearing an overcoat. They devise a contest to see who can get the coat off the man first. The wind blows hards and almost whisks the coat right off the man's back, but the man grabs it and hugs it tightly around him. The wind swirls around, huffs and puffs some more, but nothing works. It is the sun's turn. She shines brightly for a few minutes and the man removes his overcoat. Governments should try and act like the sun in this tale when trying to encourage energy efficiency savings as part of a global warming strategy. They should create energy pricing structures which encourage efficiency and environmental protection, provide clear and simple information, as well as making it positively desirable through effective marketing for the public to take the energy efficiency route. This approach, rather than the negative image of 'Save It' campaigns, would help to unleash the potential of the general public to achieve the difficult target reductions in carbon dioxide emissions we now have to face.

Machiavelli once said, 'there is nothing more difficult to carry out . . . than to initiate a new order of things. For the reformer has enemies in all who profit by the old order, and only lukewarm defenders in all those who would profit by the new order.' Robert Malpas, who uses this reference in some of his speeches, says that 'change is a function of *dissatisfaction*, *vision* and *practical first steps*.' Each of the actions in our 'Check List for Action' (pp. 216–217) is a practical step towards the goal of staving off major climatic change.

Policy action on the supply side

If energy efficiency is the main 'cutting edge' of our energy policies in the immediate future, we also need to assess new energy supply options, choosing substitutes for our current use of fossil fuel. Natural gas produces only 60 per cent of the carbon dioxide emissions of coal for the same amount of energy, and 75 per cent of that from oil. Until recently, gas was treated as a premium fuel in many countries, especially in Western Europe. Due to abundant new supplies from Siberia, the North Sea and North America, however, such views are changing. Reserves are now expected to last from forty to eighty years, even with a growth in use. If burned in efficient power-stations and boilers, natural gas has an important bridging role, buying us time to switch to carbon-free sources. Its use would be stimulated through a 'carbon tax' which was higher for coal and lower for gas.

Carbon-free supply sources currently available are essentially nuclear power and renewable energy. We've discussed the role of nuclear in the last chapter and discovered its limitations and environmental problems. Chapter 5 assessed some of the problems and limitations of renewable energy, which clearly offers no simple solution to our problems, and will require further development of a number of technologies and changes in our institutions. However, in the view of the Toronto Conference and Hamburg Congress, it will provide the main alternative source of energy to fossil fuels.

The incoming solar energy absorbed by the Earth in one year is equivalent to fifteen to twenty times the energy stored in all of the world's reserves of recoverable fossil fuels. If just 0.005 per cent of this could be captured with fuel crops, specially designed buildings, wind and water turbines, solar collectors and wave energy converters (see appendix 4) it would supply more useful energy than we currently get from burning oil, coal and gas. Renewable energy currently provides 21 per cent of the world's energy – 15 per cent from biomass and 6 per cent from hydro-electric power. Its potential contribution is very much larger. Christopher Flavin and Cynthia Pollock Shea of the Worldwatch Institute have suggested that a four- to five-fold increase over fifty years is realistic.

The progress of renewables in some countries has been impressive. Brazil, Israel, Japan, the Philippines, Sweden and Denmark are well on their way towards a major reliance on renewable sources. Twenty countries are producing power from natural geothermal steam, and installed capacity is expected to reach 8,000 megawatts by 1992. The USA already gets 4 per cent of its energy from biomass, more than is generated from nuclear power, and has 1,500 megawatts of wind power, 500 megawatts of this being installed in a single year. Brazil gets 60 per cent of its energy from renewables, mainly hydro and biomass. Over 4.5 million biomass plants in China produce over one billion cubic metres of gas per year. Around 10,000 small straw burning furnaces on farms in the UK save the equivalent of 100,000 tonnes of coal a year. In Japan, over 4 million solar water heaters are in use. In Israel, the figure is 700,000.

The future potential is large. Research for the European Community has suggested that more than 400,000 sites would be suitable for large wind turbines. These can provide more than three times the community's current electricity use. The UK has one of the best wind and tidal machines in the world. Each of these sources can provide some 20 to 30 per cent of current electricity demand, at prices competitive with current fuels. Through its very diversity – renewables can provide heat, electricity and transport fuels – it allows greater flexibility from a village to city scale, without the risks of the nuclear option. To prepare for a 'solar age' in the next century, there is much we need to do now.

Futuristic energy sources

More futuristic energy resources which might make a contribution in years to come include hydrogen and nuclear fusion. Both suffer from poor economics and technical uncertainties, but could have a role to play by the middle of the next century. Fusion is perhaps the furthest away – some scientists think seventy to a hundred years before commercial viability could be proven, if ever. It will thus be too late to reduce global warming.

Fusion research has already cost $8,000,000,000 in Europe alone, and at least $20,000,000,000 would be needed over the next forty years to reach technical and commercial viability. Fusion also produces radioactive-waste.

Hydrogen fuel isn't an energy source as such – it is a secondary energy carrier produced by electricity acting on water. If the electricity came from coal, the exercise would clearly be a pointless one, so nuclear power or renewables are the preferred options. Cars running on hydrogen have been developed, but the whole system is at least a factor of six away from commercial viability.

Laying the institutional framework

Renewables are still regarded by many countries as a 'Cinderella-option', hovering on the edge of economic viability, but without the institutional and political support to survive during periods of low fossil fuel prices or budget cuts. Such budget cuts occurred in the UK in 1982, when a promising wave-power programme was slashed. Ronald Reagan slashed solar energy research budgets by two-thirds in less than four years in the mid-1980s. Kai Millyard of FOE Canada has criticised his government's cuts in the energy conservation and renewable energy programmes from \$140 million (Canadian) in 1984 to roughly \$74 million in 1988. Total renewable energy research funding throughout International Energy Agency (IEA) member countries has fallen by 64 per cent from its 1980 peak.

The need to remove hidden subsidies from fossil fuels and nuclear, and to include their true environmental costs is a crucial element in allowing renewables any success in the market place. Michael Flood has shown that the UK nuclear industry has had a subsidy of over £13 billion since the late 1950s, which equates to over one pence for 'every single kilowatt of electricity the industry has ever generated'. The pattern is similar in the USA, France and the USSR.

Sustained research and development programmes, linked to well-supported demonstration schemes, and a programme of good public and commercial information, would help immensely in improving the efficiencies of technologies such as solar cells and wind generators. It would also cut costs and take away some of the inherent risks of a new technology.

Action check list

☐ Set up well-resourced renewable energy and energy conservation agencies to co-ordinate research, help fund development projects, and provide practical information for business and the general public.
☐ Set targets for the utilisation of renewable sources. Most countries could be getting at least 20 per cent of their energy needs from renewables by the year 2020.
☐ Introduce a range of financial incentives for renewables which reflect their role in reducing greenhouse-gas emissions.
☐ Make sure building codes and regulations encourage the construction of passive solar housing.

Your Personal Check List

ENERGY
- INSULATE THE LOFT, WALLS AND WATER TANK
- FIT DRAUGHT-STRIPPING TO DOORS AND WINDOWS ☐
- USE EFFICIENT LIGHT BULBS & APPLIANCES ① ☐
- FIT RADIATOR THERMOSTATS ☐
- FIT A GAS CONDENSING BOILER ☐

PAPER/TREES
- USE RECYCLED PAPER
- AVOID DISPOSABLE PRODUCTS - REFUSE EXCESS PACKAGING ☐
- PLANT SOME TREES ☐
- AVOID TROPICAL HARDWOODS WITHOUT A "GOOD WOOD" SEAL OF APPROVAL ② ☐

CONSUMER POWER
- BUY LOCALLY PRODUCED GOODS
- BUY ONLY CFC-FREE AEROSOLS ② ☐
- BOYCOTT COMPANIES WHO CUT DOWN TROPICAL RAINFORESTS ② ☐

TRANSPORT
- USE PUBLIC TRANSPORT
- SHARE YOUR CAR ☐
- CHOSE A FUEL EFFICIENT, LOW POLLUTING CAR ⑤ ☐
- USE YOUR BICYCLE FOR SHORT TRIPS ☐

CAMPAIGNING
- JOIN AN ENVIRONMENTAL GROUP
- WRITE LETTERS TO YOUR M.P. AND LOCAL NEWSPAPER ☐
- GET YOUR LOCAL COUNCIL TO PLANT MORE TREES AND SAVE ENERGY ☐
- BUY A COPY OF THIS BOOK FOR A FRIEND ☐

FOOD
- EAT LESS MEAT
- BUY ORGANIC - AVOID PESTICIDES ☐

① SEE THE 'GREEN CONSUMER GUIDE'
② DETAILS FROM FRIENDS OF THE EARTH
⑤ DETAILS FROM GREENPEACE

Fig. 16

Methane and other greenhouse gases

Action to reduce the remaining greenhouse gases is extremely varied. Reductions in fossil fuel use and the cleaning up of emissions will help to significantly cut the formation of tropospheric ozone and nitrous oxide. Reduced use of nitrogen-based fertilisers would help to reduce nitrous oxide formation. This would also help to reduce agricultural surpluses and water pollution. This can be achieved by introducing nitrogen quotas, and grants for less-intensive farming methods. Consumers can help by insisting on organic food.

Methane is a troublesome greenhouse gas. As chapter 2 indicated, its sources are both varied and uncertain. George Woodwell of Woods Hole Institute in Massachusetts believes that we need to solve the carbon dioxide problem in order to stop methane levels increasing. Woodwell argues that it is the warming trend that has already set in which is stimulating the increasing decay of organic matter and the release of methane from the tundra. James Lovelock points the finger at cattle: 'If it were possible to vastly diminish the number of cattle world-wide it would be one of the most useful single things that humans world-wide could take.' He argues that being a vegetarian is an important step in reducing global warming, by reducing the demand for beef.

Significant amounts of methane come from waste burial sites, particularly in the industrialised countries which dispose of huge quantities of waste. The US throws away 16 billion 'disposable' nappies alone each year. Britain dumps 30 million tonnes of degradable waste every year in 5,110 waste disposal sites, 2,100 of which are capable of generating methane. Methane can be extracted relatively easily as landfill gas and burned to generate electricity or produce heat. It has been estimated that the methane seeping from landfill sites is equivalent to 50 to 60 million tonnes of carbon dioxide, 10 per cent of the total UK emissions. Incinerating the waste instead would produce about 16 million tonnes of carbon dioxide, but save around 6 million tonnes of coal and 16 million tonnes of carbon dioxide if the heat were used. Less packaging and more recycling of paper would help to reduce some methane gas production at source.

Last, but not least is the leakage of methane from natural gas pipes and installations. Greater attention to this through better housekeeping and more frequent tests would help reduce losses, as would a banning of gas appliance pilot lights which often blow out and waste gas.

Action check list

☐ Choose cleaner and more energy efficient cars.
☐ Governments should bring in vehicle emission standards at least up to US standards.
☐ Introduce nitrogen quotas in farming.
☐ Legislate for recycling schemes.
☐ Assess methane production from waste-landfill sites and encourage the development of landfill gas plants and tightly regulated incineration schemes.
☐ Introduce a code of practice in the gas industry for reducing gas leaks, and ban pilot lights in gas appliances through regulation.

Adapt or die

The major emphasis of this chapter has been on ways to substantially reduce greenhouse-gas emissions. We discovered in chapter 2 that some climate change is already underway. On top of the 0.5°C temperature increase already with us, the Earth is already committed to a further increase of 0.5 to 1°C, and there are likely to be changes in sea levels, temperature and weather patterns as a result. Some form of adaptation, which is the last component of our four-part strategy, will therefore be necessary. Low-lying coastal regions, sensitive eco-systems, and regions already vulnerable to fluctuations in water supply are priority action areas.

Action check list

☐ Assess coastal regions for potential sea-level rise and increased storm damage. Coastal management plans should be developed, including the building of new sea-walls.

☐ Consider abandoning farmland in some areas. Governments should consider setting up compensation or 'buy out' schemes for vulnerable areas.

☐ Develop and maintain 'migration' corridors for eco-systems likely to substantially shift. These need to be considered where new road, agricultural and housing developments are planned. -

☐ Water reservoirs and flood-control systems should be assessed for their possible extension.

☐ Food reserves may have to be increased for countries likely to suffer increased drought or unpredictable weather in future.

☐ Re-assess farm practices and seed strains in order to develop a more resistant and adaptable system, even where some loss in output might occur.

☐ Improve food distribution both nationally and inter-nationally to cope with potential food shortages, as well as international mechanisms and funds to cope with disaster relief, famine and increased debt.

International agreements

Global warming is truly international in scale, even though the root causes and actions to reduce it are to be found on an individual and local scale. Important though individual action is, it would be difficult, as George Woodwell has said, 'for an individual to live out of the context of his or her time', particularly if this would make living much more difficult and expensive.

As chapter 5 argued, international co-operation and agreements are essential if we are to move to a world where all countries adhere to policies which will hold down greenhouse-gas emissions. China has an official policy of doubling coal consumption within fifteen years. Even if all the other countries in the world agreed to reduce carbon dioxide and other greenhouse-gas emissions, this would be swallowed up if China achieves her goal. Freely negotiated but binding agreements are essential to prevent this.

Negotiations for greenhouse-gas reductions are going to be difficult to conclude. They will have to make allowances for varying levels of development in different countries, and the efficiency with which they currently use fossil fuels. Denmark and Japan, for example, are using energy much more efficiently than the USA and Britain – they should therefore have lower targets for carbon dioxide reductions. One suggestion by Michael Grubb, from the Royal Institute for International Affairs in London, is to give each country greenhouse gas 'production shares' on the basis of current 'energy per capita' use. Shares could then be traded internationally, allowing some increases in the developing countries, in return for 'technology aid' and reductions by the developed world. Another option is to allocate carbon dioxide 'quotas' and then set reductions for each country. This could be a long drawn-out set of negotiations however, with over 100 separate 'quotas' to agree.

Bill Chandler, of Batelle Laboratories in Washington, favours an energy efficiency Protocol where annual improvements in efficiency are set in order to achieve a global emissions reduction goal. He points out that a 3.8 per cent per annum improvement, not much above the level achieved in OECD countries between 1973 and

1985, would achieve the Toronto Conference goal of a 20 per cent reduction by the year 2005.

It is beyond the scope of this book to look at the range of possibilities under international agreements. Although we should take heart from the speed and relative goodwill in which the Montreal Protocol was negotiated to protect the ozone layer, we should have no illusions about the difficulties of agreeing a Carbon Dioxide or Methane Protocol. The public, the media and non-governmental organisations therefore have an important role in pressurising governments, publicising the issue, and pushing for both unilateral and multilateral action by businesses, communities and nations. Only public pressure will ensure that governments respond in time.

A low risk policy

Inaction is not a neutral, low-risk policy, but rather a gamble that risks much greater harm.

Gus Speth, President of the World Resources Institute, 1986.

Faced with a 50:50 chance of winning on a horse race or a lottery, many would be tempted to wager a few pounds. If the consequences of losing the bet were to lose your family and home, there would be few takers. The consequences of humanity's 'uncontrolled, globally pervasive' experiment with the Earth's atmosphere are 'a major threat to international security and are already having harmful consequences over many parts of the globe'. Faced with the full facts of global warming, and given that the odds of there being any 'winners' in such an unpredictable climatic future are considerably worse than 50:50, few individuals and countries are likely to place bets. The sheer magnitude of the risks from global warming make a 'wait-and-see' policy intolerable,

The Four-Track Strategy we have outlined is a 'low-risk' strategy, in contrast to the wait-and-see policy which, according to Wilfrid Bach from the University of Munster, 'is one of high risk and may very well be deadly.' It is an approach which advances basic research as well as research on the causes and impacts of global warming, while taking precautionary measures.

On 2 February 1989, Senator Tim Wirth introduced a bill called the Natural Energy Policy Act to the US Senate in an attempt to stimulate specific action by the US government to 'reduce the generation of carbon dioxide and trace gases as quickly as is feasible'. When introducing his bill, Wirth described the Earth's history as if compressed into just a hundred years:

> The dinosaurs came and left about one year ago. Man arrived only two weeks ago. We began the widespread use of fossil fuels at the start of the industrial revolution only five minutes ago. But the legacy of those five minutes is large. In these brief five minutes we have upset more than ninety-nine years of development of the Earth's environment. If we are to protect the predictability of our climate for future generations, we must act in the next four seconds.

8

TOWARDS 1992 AND BEYOND

As we near the end of the twentieth century, humanity faces
a crucial question: will we devote our abilities, our energy,
and our efforts to further short-term material well-being, or
will we commit ourselves to enhancing life on planet Earth?
Gro Harlem Brundtland, June 1988.

No one who has objectively assessed the pressures and
damage human beings are inflicting on the environment
can be under any illusions about the challenge of 'enhancing life on planet Earth.' Our short exploration of the
world's atmosphere and climate has revealed the delicate
balance between the Earth's systems which allow us to
exist and prosper. As long ago as 1965, Adlai Stevenson
described our journey as 'passengers in a little spaceship,
dependent upon its vulnerable reserves of air and soil;
committed for our safety to its security and peace; preserved from annihilation only by the care, the work, and
I will say, the love we give our fragile craft.'

We have travelled a long way since then. Global environmental summit meetings are now the order of the
day, membership of environmental organisations has
grown rapidly, and scientific knowledge of the Earth's
systems has increased dramatically. NASA's 'Mission to
Planet Earth' is a tacit admission that in our rush to
explore new planets we have neglected our own. 'Mission to Planet Earth' is also a suitable metaphor for the

new awareness of our planet amongst scientists, politicians and growing numbers of ordinary people.

The task of averting global climate change has been made more difficult by the increasing levels of toxins and gases we spew out from our homes, factories, farms and vehicles. Since Adlai Stevenson's famous speech, carbon dioxide emissions have increased by 35 per cent to over 6 billion tonnes a year. Yet our understanding of the severity of the problem has barely permeated the meeting rooms of policy-makers. Global warming is already upon us. As Ronald Sagdeev, Director of the Soviet Space Research Institute has admitted, 'the crime is already committed, but the final sentence may be softened if we are clever enough.' Cleverness will not be enough. Human beings have achieved extraordinary things in the short time we have been on the planet; we now face our greatest challenge – the will to survive.

The will to survive is already starting to break down entrenched political ideologies. The arid cold war is being replaced by a different type of war, not a nationalistic or a military campaign but a universal crusade to save the planet.

Leadership

> Example moves the world more than doctrine.
> Henry Miller, *The Cosmological Eye*

At a 30 November 1988 breakfast meeting with US President-elect Bush, environmental leaders representing 6 million members presented their detailed report *Blueprint for the Environment*. Borrowing from a Bush campaign slogan, National Wildlife Federation President Jay Hair told Bush: 'Read my lips: protect the environment.' Bush responded, 'I will.'

Our will and ability to survive depends heavily on the

leadership of the countries such as Japan, the Soviet Union and especially USA – the greatest consuming and polluting country in the world. President Bush has so far made positive noises, admitting that '1988 was the year the Earth spoke back'. He has committed himself to 'integrating environmental considerations into all policy decisions' and to a global environmental summit in 1990. Competition between leaders like Bush, Gorbachev and Thatcher for pre-eminence in the 'green leader' stakes is no bad thing in itself. However, environmental rhetoric, conferences and publications must be backed up with specific actions – and fast. It is on action not words that people must assess their leaders and pass judgement. UNEP Director Mostafa Tolba has set the pace by calling for a Global Climate Convention by 1992. Politicians must respond to this challenge.

Many of the actions needed from our leaders, businesses and others are outlined in chapter 7, and have been argued for at the Toronto and Hamburg conferences. As the Hamburg Congress concluded, tackling the problem of global warming will require 'an extraordinary level of organisation, leadership, and responsibility from politicians, industry and individuals'.

Ten years to save the planet

Conditions are never just right. People who delay action until all factors are favourable are the kind who do nothing.
William Feather, *The Business of Life*

Scientists have a vital role in the process of changing policies in order to reduce greenhouse gases. It is important that they retain values of objectivity and truth when evaluating complex problems. Yet, as issues such as acid rain, leaded petrol and low-level radiation have shown, the opportunity provided by scientific uncertainty for political delay is immense. It may be years before scientific evidence satisfies some scientists and politicians. Peter Chester, Environment Director for the Central Electricity Generating Board in Britain, regards the science of climatic change as one that is 'only just cutting its teeth' and feels that the first priority is 'intensified and focused research so that hard decisions can be made on the basis of hard fact.' His estimate for this process is ten years. Robert Goodlad, the World Bank's environment chief for Latin America, argues that the world has 'ten years remaining' to save itself. Ram Sundararaman of the WMO agrees and insists that 'policies that have real teeth to them ... have to be in place five or ten years from now.' John Houghton, head of the scientific IPCC committee advising governments, sees the task of scientists to 'quantify uncertainty in a useful way, which will be readable and understandable by people who have to make policy ... without being too cautious.'

The need both for certainty and swift action puts scientists in a difficult dilemma admits Robert Watson of NASA. He suggests, however, that 'the more frivolous the utilisation of a chemical such as CFCs, or the process producing other greenhouse gases, the less scientific certainty I believe we need.' Michael Oppenheimer, Environmental Defense Fund's senior scientist, goes further: 'It's time to move on this issue ... there is no advantage in waiting. If we don't move fast, there will be so much warming that our policy options will be narrowed in future.' Watson, Oppenheimer, Woods Hole Institute Director George Woodwell, and Sherry Rowland, one of the pioneers on the ozone depletion issue, have been prominent in a group of scientists calling for action now. Where options such as energy efficiency and halting tropical rainforest destruction make sense anyway, and would solve other environmental problems, the case for immediate action seems irrefutable.

Although scientists and politicians have important roles to play in combating climatic change, industry too has a crucial role, particularly as some multinational companies have greater power and resources than many sovereign nations. Their corporate decisions on issues such as energy technologies, the buying up of tropical rainforests for cattle ranching, and the use of CFCs and their substitutes, all affect the planet.

Welcome signs of more environmentally-inclined thinking are being seen in a few company boardrooms. IBM, for example, is trying to live up to a 'sustainable development' policy within its own operations. It already donates more to UNEP than the UK government, and has a corporate policy to do more than is legally required to protect the environment. In West Germany, pollution control manufacturers have responded quickly and in an innovative way to strict standards for air and water pollution. They now have a turnover of more than 20 billion DM per year. In the USA, the 3M company, through its 'Pollution Prevention Programme' has so far saved $420 million as well as a great deal of pollution. Bejam, a major food and fridge/freezer retailer in the UK, has just commenced taking their customers' old fridges and recycling the CFCs with the co-operation of ICI. The UK Chemical Industries Association, which includes a major CFC producer (ICI) and a number of CFC users, launched a 'responsible care' programme in late 1988. Observers will need time to be fully convinced by this, but the Association has at least formally admitted that 'rather than pretending to be ecologically neutral, we must acknowledge that much of what the industry does changes the natural order of things'.

Individual action

Out of action, action of any sort, there grows a peculiar useful, everyday wisdom.

Dr Frank Crane, *Habit* (Essays)

Biosphere II is based in the desert of Arizona, USA. Eight humans will seal themselves into a giant greenhouse for two years, with 4,000 plant and animal species, and a selection of man-made ecosystems, to see if they can all survive. Sunlight will be the only outside energy source allowed in. The experiment is an attempt to see whether the ecosystems will evolve and deal with the pollutants and emissions, and humans can survive in a manufactured system. The rest of us will have to be content with our continuing experiment on Biosphere I – planet Earth.

Unless we want to turn our back on the planet, we will all have to take individual responsibility and make a stand. We are all consumers, members of families and organisations, voters and shareholders – with the ability to make choices and to change the world. Human history is littered with ordinary men and women who did extraordinary things, changed the views of others, and influenced nations. To look for leadership is a natural instinct, but to abrogate individual responsibility is to give up a part of the human spirit. Former British diplomat Ronald Higgins wrote that in comparison with the world's serious environmental problems 'apathy is the greatest enemy of all.' Edmund Burke once said, 'nobody made a greater mistake than he who did nothing because he could do only a little.' We all have to do a little to solve the world's big problems. As Mostafa Tolba of UNEP puts it: 'we need small-scale change, but on a widespread basis.'

The Earth's systems are complex, and yet beautifully simple. The interlinking of processes and the compensating nature of the planet, in the face of both man-made and natural changes, can appear almost pre-ordained.

The Gaian philosophy embraces this concept. Much remains to be proven about this hypothesis, but Gaia can teach us a new respect for the planet, for its power and its secrets.

The inter-relationships of problems and solutions to climatic change has been a theme of this book. The dual role of CFCs and of acid emissions has been described. The synergistic effect of multiple pollution assaults on ecosystems may lead to a threshold of change much sooner than the scientific models predict. Solutions need to take account of such uncertainties.

In the race to save our planet, a number of concepts regarded by policy makers as accepted wisdom need to be overturned. The rejection of energy consumption as a symbol of economic development and growth is an important hurdle for many countries. It needs to be replaced by an approach which places a cost on environmental pollution, and is based on the premise of completing a task at the lowest overall cost to the economy and the environment. A high-energy growth route is not an historical inevitability. High-energy consumption, particularly fossil fuels, would bring in its wake further degradation of the planetary systems and slow the necessary development of three-fifths of the inhabitants of the world. Energy efficiency and the increasing use of renewable energy sources, have the potential to allow speedier and more sustainable economic development, without the political and social problems of nuclear power. Energy efficiency is one of the strongest and least divisive weapons we have in our 'universal crusade to save the planet', and brings multiple benefits. It needs to be at the top of every agenda on climatic change.

It is to you, as an individual, that this book is addressed. As *Our Common Future* states 'first and foremost our message is directed towards people, whose well-being is the ultimate goal of all environment and development policies.' The role of individuals has been highlighted on many occasions in this book. Although we agree with George Woodwell that it is 'difficult for individuals to act alone, out of the context of his or her time', our behaviour is the root cause of global warming. We need to seek out that root cause in its many manifestations.

Root causes

Ignorance is an important root cause – and one which governments, the media, non-governmental organisations, teachers and businesses can do much to end. Ignorance about the impacts of the products we buy, the transport systems we choose and other aspects of our lifestyles, can be substantially remedied through simple information, product labelling, and higher standards enforced through regulation.

Poverty and debt result in causes which lead to environmental destruction and short-term planning. The appalling imbalance between resources in the developing nations and the developed world, and the huge net outflow of money from developing countries cannot continue indefinitely. The $35 billion the industrialised world spends on development assistance must be measured against the more than one trillion dollars the developing world owes it in loans and interest payments. Debt is leading to the pressures and short-term horizons which are stripping forests and impoverishing agricultural land. These in turn threaten the whole planet – 'our common future.'

Poverty and greed lie behind the destruction of the Amazonian rainforest. The murder of ecologist and tribal activist Chico Mendes in 1988, because of his opposition to cattle-ranchers and others destroying the forest, was closely followed by a statement from Ronaldo Caiado, president of a landowners' organisation. 'The Amazon is ours to do what we like with,' said Caiado, 'we have the right to clear it and grow food to feed Brazilians.' Even if the cattle reared on cleared forest land were actually going to feed Brazilians, which they are not, the short-term attitude is revealing. Within five to ten years the cleared rainforests will become depleted in nutrients, and virtually useless for crops or cattle.

Seeds of hope

After several decades of poorly designed and ill-managed aid projects, aid donors and recipient countries have now started to understand the basis of successful projects. They must fulfil real rather than perceived needs; they must involve the people who must make the project work in the long term, in an open and participatory way; they must reflect local conditions such as social customs and topography; and they must be flexible. Large, inflexible schemes brought in by outside 'experts' often fail. Some recent success stories include:

☐ **Haiti** – the involvement of non-governmental organisations (NGOs) in an Agroforestry Outreach Project has led to the provision of 25 million tree seedlings for 110,000 farmers and 39 nurseries producing 3 million seedlings a year.

☐ **Kenya** – the development of more advanced cooking stoves (jikos) has doubled their efficiency to 40 to 45 per cent since 1981, reducing charcoal demand by 1.5 million tonnes. Local craftsmen amended original designs from Western experts which proved unreliable, and now advise other African countries on setting up stove manufacturing facilities.

☐ **Zimbabwe** – although listed as one of twenty drought-stricken African countries in 1985, it regained self-sufficiency in maize through a large increase in output from small farmers. Attractive government pricing systems, easier maize collection and more accessible credit facilities all contributed to this success.

☐ **Philippines** – the Asian Development Bank has approved $420 million for reforesting about 360,000 hectares of denuded and open forest lands. Many of the 185 reforestation projects will be coordinated by NGOs rather than government officials due to their greater knowledge and success in local areas.

☐ **India** – the concept of 'integrated village energy systems' is now a reality, with well over forty villages using biogas, wind, solar, photovoltaic and other renewable energy and many more systems now planned. Though not without problems, many of the villages are generating much more employment and income than before, without the economic and environmental problems of relying on fossil fuels. Participating villages are expected to grow to more than 200 within the next few years.

4,000 days

There are about 4,000 days left in this century. But the scale of the challenges before us mean that we shall all need to work, in one way or another, on every day of the coming decade if we are to meet these challenges.

> Mostafa Tolba, at the opening session of the
> Inter-governmental Panel on Climate Change,
> 9 November, 1988.

On the day the manuscript for this book was handed to our publisher, there were 3,945 days left in this century. By the time you buy the book, yet more days and weeks will have slipped away. Each week more than 500,000 new cars roll off the production lines around the world, ready to pollute the air. Each day, more than 50 million tonnes of carbon dioxide from the burning of fossil fuels, plus millions of tonnes of other greenhouse gases, will have escaped to the atmosphere. The destruction of the tropical forests, home to at least half the Earth's plant and animal species, will have continued at a rate equal to one football field a second.

Taken on a global scale, the seemingly relentless growth of pollution, population and loss of farmland can induce despair and hopelessness. Yet many countries have immense reserves of natural resources and wealth. Human ingenuity, which is capable of dealing with the root causes of these problems, has been tested before, often in war unfortunately, or sometimes in the face of technological challenges such as the conquest of space or biotechnology. It has and can respond successfully.

As individuals, few of us would pour poisons into our own backyard or deliberately upset local weather patterns. We tend to be very protective about our own patches of greenery. Yet we somehow manage to divorce ourselves from the consequences of our lifestyles and actions when they are out of sight. Climatologist Hubert Lamb has observed that:

> continual efforts and education are needed to guard against mistaken actions and policies, which are liable to arise from modern urban societies' increasing remoteness from and unacquaintance with the natural world.

Modern industrial society has been described by Brazilian ecologist Jose Lutzenberger as 'a fanatical religion' with its seeming obsessions with ever-continuing growth. Nobody plans to alter world climate, to change the seasons or to make countries destitute by removing huge areas of forests – it just happens doesn't it? Things don't just happen, however. In an age of mass communications, with powerful and cheap computers available to millions, and the instant access to events by TV reporters, our apparent lack of comprehension and understanding of the connection between our actions and the results can no longer be excused. It is our duty to under-

Global Cooperation

" IT IS TIME THAT WE REALISED THAT
WE ALL SHARE A COMMON FUTURE "
GRO HARLEM BRUNDTLAND, PRIME MINISTER OF NORWAY, JUNE 1988

① **USA – GUATEMALA**
PLANTING TREES TO SOAK UP CARBON DIOXIDE

② **EEC – SAHEL**
SUPPORT FOR SMALL FARMS

③ **U.K.–CHINA**
TECHNOLOGY TRANSFER OF CFC SUBSTITUTES

④ **USA – USSR**
SCIENTIFIC EXCHANGES ON OZONE DEPLETION & ENERGY EFFICIENCY

⑤ **AUSTRALASIA – ANTARCTICA**
PROTECTING THE LAST WILDERNESS

⑥ **NEW YORK – U.N.**

⑦ **NAIROBI – U.N.E.P.**

⑧ **CANADA – PHILIPPINES**
SUPPORT FOR WOMENS ENVIRONMENTAL GROUPS

⑨ **SCANDINAVIA – SOUTH AMERICA**
WRITING OFF CRIPPLING DEBTS

" THE DOZEN YEARS BETWEEN NOW AND THE END OF THE
CENTURY FORM A WINDOW OF OPPORTUNITY FOR GLOBAL
COOPERATION… LET US SEIZE THAT OPPORTUNITY "

MOSTAFA TOLBA, DIRECTOR GENERAL OF U.N.E.P., NOVEMBER 1988

Fig. 17

stand the natural world, develop a new cooperative relationship with it, and keep it fit to live in. The root cause of global warming is ourselves. The solutions also lie with us.

APPENDIX 1

Statement from Villach and Bellagio Workshops: Developing policies for responding to climatic change

A summary of the discussions and recommendations of the workshops held in Villach (28 Sept–2 Oct 1987) and Bellagio (9–13 Nov 1987), under the auspices of the Beijer Institute, Stockholm.

1. The atmospheric concentrations of a number of trace gases are increasing as a result of human activities. These gases have an important effect in trapping energy at the Earth's surface and in the lower atmosphere (the 'greenhouse effect') leading to a warming thus to changes of climate.

2. It is now generally agreed that if the present trends of greenhouse gas (GHG) emissions continue during the next hundred years, a rise of global mean temperature could occur that is larger than any experienced in human history.

3. A two stage workshop process held in Villach (Austria) and Bellagio (Italy) in 1987 examined how climatic change resulting from increasing GHG concentrations could affect environment and society during the next century and explored the policy steps that should be considered for implementation in the near term.

4. Scenarios of global climatic change that could occur between now and the end of the next century as a result of continuing emissions of GHGs were developed. The upper bound scenario, which considers a large increase of GHG emissions and a high sensitivity of the climatic response, gives a global surface temperature increase of 0.8°C per decade from the present until the middle of the next century. The middle scenario, which considers current

trends in GHG emissions, a reduction of chlorofluoro-
carbon emissions according to the Montreal Ozone Protocol, and a moderate climate sensitivity, gives a temperature
increase of 0.3°C per decade. The lower bound scenario,
which assumes a strong global effort to reduce GHG emissions and relatively low climate sensitivity, gives a rate of
temperature increase of 0.06°C per decade.

5. The most extreme temperature increases would probably
occur during winter in the high latitudes of the Northern
Hemisphere, where the changes could be two to two and a
half times greater and faster than the globally averaged
annual values. Precipitation changes could include enhanced winter precipitation in the high latitudes, and,
perhaps, a decrease in summer rainfall in the mid-latitudes.

6. GHG-induced global warming could accelerate the present
sea-level rise, probably giving a rise of about 30cm but
possibly as much as 1.5m by the middle of the next century.
The effects of this will include: erosion of beaches and
coastal margins; land-use changes; wetland loss; increased
frequency and severity of flooding; damage to port facilities, coastal structures and water management systems.

7. In the middle latitudes the main impacts are expected to be
on the relatively unmanaged ecosystems, in particular the
forests. If the temperature change is rapid, dieback of trees
will result and more and more forest would need managing
to maintain it in a productive mode. For the lower bound
scenario of temperature change, extinction of species, re-
productive failure and large-scale forest dieback would not
occur before the year 2100. A further effect could be the
release of a significant amount of carbon from soils, trees
and other plants as carbon dioxide and methane and this
would enhance the greenhouse warming.

8. Climatic change will not occur in isolation. Increasing
amounts of atmospheric and aquatic pollutants can be
expected from urban-industrial growth. The response to
climatic change will be affected by these pollutants. The
importance of these interactions and the need to investigate
them further cannot be overestimated.

9. In the semi-arid tropical regions the climatic changes that
might occur by the middle of the next century as a result of
the increasing concentrations of GHGs include a temperature increase of the order of 0.3 to 5°C and a decrease in

precipitation rate in one or more seasons. These changes could worsen the current critical problems of the semi-arid tropics, especially through their effects on food, water and fuelwood availability, human settlement patterns and the unmanaged ecosystems.

10. In the humid tropical regions it is expected that the GHG-induced changes could include a warming of 0.3 to 5 °C and an increase in rainfall amount. In addition, tropical storms might extend into regions where they are now less common. Coastal and river regions and region of infertile soils in the uplands appear to be especially vulnerable.

11. In the high-latitude regions it is expected that the mean winter temperature could increase by between 0.8 to considerably more than 5 °C by the middle of the next century. In addition, there could be a withdrawal of the summer pack ice, increased cloudiness and precipitation, slow disappearance of the permafrost and changes in the tundra and in the northern limit of the boreal forest. These changes can be expected to affect marine transportation, energy development, marine fisheries, agriculture, human settlement, northern ecosystems, carbon emissions, air pollutions, and security.

12. The rate of global temperature change that would occur if current trends of GHG emissions were to continue are large compared with observed historic changes and would have major effects on ecosystems and society. For this reason a co-ordinated international response will become inevitable.

13. Adaptation strategies for responding to a changing climate adjust the environment or our ways of using it to reduce the consequences of a changing climate; limitation strategies control or stop the growth of GHG concentrations and limit the climatic change. A prudent response to climatic change would consider limitation *and* adaptation strategies.

14. Whatever limits on climatic change might be implemented, planning and decision-making could be facilitated by the use of long-term environmental targets, such as the rate of temperature or sea-level change. The choice of a target would be based on observed historic rates of change that did not put stress on the environment or society. The environmental target can be translated into emissions targets for GHGs that could be used for regulatory purposes.

15. An evaluation of the changes of GHG emissions that would be required to limit the global warming rate to the largest natural rates of increase observed in the last century, suggests that the limitation could only be accomplished with significant reductions in fossil fuel use.

16. Strategies for adapting to or limiting climatic change could involve high costs to global society. For policy-making purposes there is a need for detailed comparisons of the costs of various strategies.

17. There are many longer-term actions that will be required in order to ensure appropriate responses to climatic changes. The actions that should receive priority now are:

   ☐ Approval and implementation of the Montreal Protocol on Substances that Deplete the Ozone Layer.

   ☐ Re-examination of long-term energy strategies with the goals of achieving high end-use efficiency.

   ☐ Intensification of development of non-fossil fuel energy systems.

   ☐ Strong support for measures to reduce deforestation and increase forested areas.

   ☐ Development and implementation of measures to limit the growth of non-CO_2 GHGs in the atmosphere.

   ☐ Identification of areas vulnerable to sea-level rise. Planning for installations near the sea should allow for the risks of sea-level rise.

   ☐ Support for and co-ordination of policy research, global monitoring activities and policy-directed scientific research on the GHG issue at the national and international levels.

   ☐ Examination by organisations, including the inter-governmental mechanism to be constituted by the WMO and UNEP in 1988, of the need for an agreement on a law of the atmosphere as a global commons or the need to move towards a convention along the lines of that developed for ozone.

   ☐ Consideration and development of the recommendations of the present Report at subsequent conferences, including the World Conference on the Changing Atmosphere (Toronto, June 1988) and the Second World Climate Conference (Spring, 1990).

APPENDIX 2

Statement issued by the participants at the World Conference on 'The Changing Atmosphere: Implications for Global Security', June 1988

1. Humanity is conducting an uncontrolled, globally pervasive experiment whose ultimate consequences could be second only to a global nuclear war. The Earth's atmosphere is being changed at an unprecedented rate by pollutants resulting from human activities, inefficient and wasteful fossil fuel use and the effects of a rapid population growth in many regions. These changes represent a major threat to international security and are already having harmful consequences over many parts of the globe.

2. Far-reaching impacts will be caused by global warming and sea level rise, which are becoming increasingly evident as a result of continued growth in atmospheric concentrations of carbon dioxide and other greenhouse gases. Other major impacts are occurring from ozone-layer depletion resulting in increased damage from ultra-violet radiation. The best predictions available indicate potentially severe economic and social dislocation for present and future generations, which will worsen international tensions and increase the risk of conflicts among and with nations. It is imperative to act now.

3. These were the major conclusions of the International Conference on 'The Changing Atmosphere: Implications for Global Security', held in Toronto 27–30 June 1988. More than 300 scientists and policy makers from 48 countries,

United Nations organisations, other international bodies and non-government organisations participated in the sessions.

4. The Conference called upon governments, the United Nations and its specialised agencies, industry, educational institutions, non-governmental organisations and individuals to take specific actions to reduce the impending crisis caused by pollution of the atmosphere. No country can tackle this problem in isolation. International co-operation in the management and monitoring of, and research on, this shared resource is essential.

5. The Conference called upon governments to work with urgency towards an Action Plan for the Protection of the Atmosphere. This should include an international framework convention, while encouraging other standard-setting agreements along the way, as well as national legislation to provide for protection of the global atmosphere. The Conference also called upon governments to establish a World Atmosphere Fund financed in part by a levy on the fossil fuel consumption of industrialised countries to mobilise a substantial part of the resources needed for these measures.

The issue

6. Continuing alteration of the global atmosphere threatens global security, the world economy, and the national environment through:
   □ climate warming, rising sea-level, altered precipitation patterns and changed frequency of climatic extremes induced by the 'heat trap' effects of greenhouse gases;
   □ depletion of the ozone layer;
   □ long-range transport of toxic chemicals and acidifying substances.

7. These changes will:
   □ imperil human health and welfare;
   □ diminish global food security, through increased soil erosion and greater shifts and uncertainties in agricultural production, particularly for many vulnerable regions;

☐ change the distribution and seasonal availability of fresh-water resources;

☐ increase political instability and the potential for international conflict;

☐ jeopardise prospects for sustainable development and reduction of poverty;

☐ accelerate extinction of animal and plant species upon which human survival depends;

☐ alter yield, productivity and biological diversity of natural and managed ecosystems, particularly forests.

8. If rapid action is not taken now by the countries of the world, these problems will become progressively more serious, more difficult to reverse, and more costly to address.

Scientific basis for concern

9. The Conference calls for urgent work on an **Action Plan for Protection of the Atmosphere**. This Action Plan, complemented by national action, should address the problems of climate warming, ozone layer depletion, long-range transport of toxic chemicals and acidification.

Climate warming

☐ There has been an observed increase of globally-averaged temperature of 0.7°C in the past century which is consistent with theoretical greenhouse gas predictions. The accelerating increase in concentrations of greenhouse gases in the atmosphere, if continued, will result in a probable rise in the mean surface temperature of the Earth of 1.5 to 4.5°C before the middle of the next century.

☐ Marked regional variations in the amount of warming are expected. For example, at high latitudes the warming may be twice the global average. Also, the warming would be accompanied by changes in the amount and distribution of rainfall and in changes in atmospheric and ocean circulation patterns. The natural variability of the atmosphere and climate will continue and be superimposed on the long-term trend, forced by human activities.

☐ If current trends continue, the rates and magnitude of climatic change in the next century may substantially exceed those experienced over the last 5,000 years. Such high rates of change would be sufficiently disruptive that no country is likely to benefit *in toto* from climatic change.

☐ The climate change will continue so long as the greenhouse gases accumulate in the atmosphere.

☐ There can be a time lag of the order of decades between the emission of gases into the atmosphere and their full manifestation in atmospheric and biological consequences. Past emissions have already committed planet earth to a significant warming.

☐ Global warming will accelerate the present sea-level rise. This will probably be of the order of 30cm but could possibly be as much as 1.5m by the middle of the next century. This could inundate low-lying coastal lands and islands, and reduce coastal water supplies by increased salt water intrusion. Many densely populated deltas and adjacent agricultural lands would be threatened. The frequency of tropical cyclones may increase and storm tracks may change with consequent devastating impacts on coastal areas and islands by floods and storm surges.

☐ Deforestation and bad agricultural practices are contributing to desertification and are reducing the biological storage of carbon dioxide, thereby contributing to the increase of this most important greenhouse gas. Deforestation and poor agricultural practices are also contributing additional greenhouse gases such as nitrous oxide and methane.

Ozone layer depletion

☐ Increased levels of damaging ultra-violet radiation as the stratospheric ozone shield thins will cause a significant rise in the occurrence of skin cancer and eye damage, and will be harmful to many biological species. Each 1 per cent decline in ozone is expected to cause a 4 to 6 per cent increase in certain kinds of skin cancer. A particular concern is the possible combined effects on unmanaged ecosystems of both increased ultraviolet radiation and climatic changes.

☐ Over the last decade, a decline of 3 per cent in the ozone layer has occurred at mid latitudes in the southern hemisphere, possibly accompanying the appearance of the Antarctic ozone hole; although there is more meteorological variability, there are indications that a smaller decline has occurred in the northern hemisphere.

☐ Changes in the ozone layer will also change climate and the circulation of the atmosphere.

Acidification

☐ In improving the quality of the air in their cities, many industrialised countries unintentionally sent increasing amounts of pollution across national boundaries in Europe and North America, contributing to the acidification of distant environments. This was manifest in growing damage to lakes, soils, plants, animals, forests and fisheries. Failure to control automobile pollution in some regions has seriously contributed to the problem. The principal damage agents are oxides of sulphur and nitrogen as well as volatile hydrocarbons. These can also corrode buildings and metallic structures, causing overall, billions of dollars of damage annually.

10. The various issues arising from pollution of Earth's atmosphere by a number of substances, are often closely interrelated, both chemically and from the point of view of potential control strategies. For example, CFCs both destroy ozone and are greenhouse gases; conservation of fossil fuels would contribute to solving both acid rain and climate change problems.

Security: economic and social concerns

11. As the UN Report on the Relationship Between Disarmament and Development states:

    > The world can either continue to pursue the arms race with characteristic vigour or move consciously and with deliberate speed toward a more stable and balanced social and economic development within a more sustainable international economic and political order. It cannot do both. It must be acknowledged that the arms race and development are in a competitive relationship, particularly in terms of resources, but also in the vital dimension of attitudes and perceptions.

    The same consideration applies to the vital issue of protecting the global atmospheric commons from the growing peril of climate change and other atmospheric changes. Unanticipated and unplanned change may well become the major non-military threat to international security and the future of the global economy.

12. There is no concern more fundamental than access to food and water. Currently inadequate levels of global food security will be most difficult to maintain into the future, given projected agricultural production levels and population and income growth rates. The climate changes envisaged will aggravate the problem of uncertainty in food security. Climate change is being induced by the already prosperous, but its effects are suffered most acutely by the poor. It is imperative for governments and the international community to sustain the agricultural and marine resource base and provide development opportunities for the poor in light of this growing environmental threat to global food security.

13. The countries of the industrially developed world are the main source of greenhouse gases and therefore bear the main responsibility to the world community for ensuring that measures are implemented to address the issues posed by climate change. At the same time, they must see that the developing nations of the world, whose problems are greatly aggravated by population growth, are assisted and

not inhibited in improving their economies and the living conditions of their citizens. This will necessitate a wide range of measures, including significant additional energy use in those countries and compensating reductions in industrialised countries. The transition to a sustainable future will require investments in energy efficiency and non-fossil energy sources. In order to ensure that these investments occur, the global community must not only halt the current net transfer of resources from developing countries, but actually reverse it. This reversal should embrace the relevant technologies involved, taking into account the implications for industry.

14. A coalition of reason is required, in particular, a rapid reduction of both North–South inequalities and East–West tensions if we are to achieve the understanding and agreements needed to secure a sustainable future for planet Earth and its inhabitants.

15. It takes a long time to develop an international consensus on complex issues such as these, to negotiate, sign, and ratify international environmental instruments and to begin to implement them. It is therefore imperative that action on serious negotiations start now.

Legal aspects

16. The first steps in developing international law and practices to address pollution of the air have already been taken: In the Trail Smelter arbitration of 1935 and 1938, Principle 21 of the 1972 Declaration of the UN Conference on the Environment, the ECE Convention on Long Range Transboundary Air Pollution and its Protocol (Helsinki, 1985) for sulphur reductions, Part XII of the Law of the Sea Convention, and the Vienna Convention for Protection of the Ozone Layer and its Montreal Protocol (1987).

17. These are important first steps and should be actively used and respected by all nations. However, there is no overall convention constituting a comprehensive international framework that can address the interrelated problems of the global atmosphere, or that is directed towards the issues of climate change.

A call for action

18. The Conference urges immediate action by governments, the United Nations and their specialised agencies, other international bodies, non-governmental organisations, industry, educational institutions and individuals to counter the ongoing degradation of the atmosphere.

19. The following actions are mostly designed to slow and eventually reverse deterioration of the atmosphere. There are also a number of strategies for adapting to changes that must be considered. These are dealt with primarily in the recommendations of the working groups.

Recommended immediate action

By Governments and Industry

20. Ratification of the Montreal Protocol on Substances that Deplete the Ozone Layer. The Protocol should be revised in 1990 to ensure nearly complete elimination of emissions of fully-halogenated CFCs by the year 2000. Additional measures to limit other ozone-destroying halocarbons should be considered.

21. In order to reduce the risks of future global warming, energy policies must be designed to reduce emissions of CO_2 and other trace gases. Stabilising atmospheric concentrations of CO_2 is an imperative goal. It is currently estimated to require reductions of more than 50 per cent from present emission levels. Energy research and development budgets must be massively directed to low and non-CO_2 emitting energy options and to studies undertaken to further refine the target reductions.

22. An initial global goal should be to reduce CO_2 emissions by approximately 20 per cent of 1988 levels by the year 2005. Clearly, the industrialised nations have a responsibility to lead the way, both through their national energy policies and their bilateral and multilateral assistance arrangements. About one-half of this reduction would be sought from energy efficiency and other conservation measures. The other half should be effected by modifications in supplies.

23. Targets for energy efficiency improvements should be directly related to reductions in CO_2 and other greenhouse gases. A challenging target would be to achieve the 10 per cent energy efficiency improvements by 2005. Improving energy efficiency is not precisely the same as reducing total carbon emissions and the detailed policies will not all be familiar ones. A detailed study of the systems implications of this target should be made. Equally, targets for energy supply should also be directly related to reductions in CO_2 and other greenhouse gases. As with efficiency, a challenging target would again be to achieve the 10 per cent energy supply improvements by 2005. A detailed study of the systems implications of this target should also be made.

24. The contributions to achieving this goal will vary from region to region; some countries have already demonstrated a capability for increasing efficiency by more than 2 per cent per year over a decade.

25. Apart from efficiency measures, the desired reduction will require: i) switching to lower CO_2 emitting fuels; ii) reviewing strategies for the implementation of renewable energy, especially advance biomass conversion technologies; iii) revising the nuclear power option, which lost credibility due to problems related to nuclear safety, radioactive wastes, and nuclear weapons proliferation. If these problems can be solved, through improved engineering designs and institutional arrangements, nuclear power could have a role to play in lowering CO_2.

26. Negotiations on ways to achieve the above-mentioned reductions should be initiated now.

27. Systems must be initiated to encourage, review and approve major new projects for energy efficiency.

28. There must be vigorous application of existing technologies to reduce: i) emissions of acidifying substances to reach the critical load that the environment can bear; ii) substances which are precursors of tropospheric ozone; and iii) other non-CO_2 greenhouse gases, in addition to gains made through reductions of fossil fuel combustion.

29. Products should be labelled to allow consumers to judge the extent and nature of contamination of the atmosphere which arises from the manufacture and use of the product.

By Member Governments of the United Nations Non-Governmental Organisations and Relevant International Bodies

30. Initiate the development of a comprehensive global convention as a framework for protocols on the protection of the atmosphere. The convention should emphasise such key elements as the free international exchange of information and support of research and monitoring, and should provide a framework for specific protocols for addressing particular issues, taking into account existing international law. This should be vigorously pursued at the international workshop on law and policy to be held in Ottawa early in 1989, the high level policy conference on Climate Change in the Netherlands in autumn 1989, the World Energy Conference, Canada in 1989 and at the Second World Climate Conference, Geneva, June 1990, with a view to having the principles and components of such a convention ready for consideration at the inter-governmental Conference on Sustainable Development in 1992. These activities should in no way impede simultaneous national, bilateral and regional actions and agreements to deal with specific problems such as acidification and greenhouse gas emissions.

31. Support the work of the Inter-governmental Panel on Climate Change to conduct continuing assessments of scientific results and initiate government to government discussion of responses and strategies.

32. Devote increasing resources to research and monitoring efforts within the World Climate Programme, the International Geosphere-Biosphere Programme and Human Response to Global Change Programme. It is particularly important to understand how climate changes on a regional scale are related to an overall global change of climate. Emphasis should also be placed on better determining the role of oceans in global heat transport and the flux of greenhouse gases.

33. Significantly increase funding for research, development and transfer of information on renewable energy, if necessary by the establishment of additional and bridging programmes; extend technology transfer with particular emphasis on the needs of the developing countries; upgrade efforts to meet obligations of the development and transfer of technology embodied in existing agreements.

34. Expand funding for more extensive technology transfer and technical co-operation projects in coastal zone protection and management.

35. Reduce deforestation and increase afforestation making use of proposals such as those in WCED's *Our Common Future*, including the establishment of a Trust Fund to provide adequate incentives to enable developing nations to manage their tropical forest resources sustainably.

36. Develop and support technical co-operation projects to allow developing nations to participate in international mitigation efforts, monitoring, research and analysis related to the changing atmosphere.

37. Ensure that this Conference Statement, the working group reports and *Atmosphere* (to be published in autumn 1988) are made available to all nations, the conferences mentioned under paragraph 30, and other future events dealing with related issues.

38. Increase funding to non-governmental organisations to allow the establishment and improvement of environmental education programmes and public awareness campaigns related to the changing atmospheres. Such programmes would aim at sharpening perception of the issues, and changing public values and behaviour with respect to the environment.

39. Financial support should be allocated for environmental education in primary and secondary schools and at universities. Consideration should be given to establishing special units in university departments for addressing the crucial issues of global change.

Toronto, June 30, 1988

APPENDIX 3

N.B. All dates in the list below are ordered month/day/ year.

Montreal Protocol Ratification Status

Country	Signed Vienna Convention	Ratified Vienna Convention	Signed Montreal Protocol	Ratified Montreal Protocol
Argentina	3/22/85		6/29/88	
Australia	*	9/16/87	6/8/88	
Austria	9/16/85	8/19/87	8/29/88	
Belgium	3/22/85	10/17/88	9/16/87	12/30/88
Burkina Faso	12/12/85		9/14/88	
Bylorussian SSR	3/22/85	6/20/86	1/22/88	10/31/88
Canada	3/22/85	6/4/86	9/16/87	6/30/88
Chile	3/22/85		6/14/88	
Congo			9/15/88	
Denmark	3/22/85	9/29/88	9/16/87	12/16/88
Egypt	3/22/85	5/9/88	9/16/87	8/2/88
Equatorial Guinea	*	8/17/88		
EEC	3/22/85	10/17/88	9/16/87	12/16/88
Finland	3/22/85	9/26/86	9/16/87	12/23/88
France	3/22/85	12/4/87	9/16/87	12/28/88
East Germany	*	1/25/89	*	1/25/89
West Germany	3/22/85	9/30/88	9/16/87	12/16/88
Ghana			9/16/87	
Greece	3/22/85	12/29/88	10/29/87	12/16/88
Guatemala	*	9/11/87		
Hungary	*	5/4/88		
Indonesia			7/21/88	
Ireland	*	9/15/88	9/15/88	12/16/88
Israel			1/14/88	
Italy	3/22/85	9/19/88	9/16/87	12/16/88
Japan	*	9/30/88	9/16/87	9/30/88
Kenya	*	11/9/88	9/16/87	11/9/88
Liechtenstein	*	2/8/89	*	2/8/89
Luxembourg	4/17/85	10/17/88	1/29/88	10/17/88
Maldives	*	4/26/88	7/12/88	
Malta	*	9/15/88	9/15/88	12/29/88

Montreal Protocol – continued

Country	Signed Vienna Convention	Ratified Vienna Convention	Signed Montreal Protocol	Ratified Montreal Protocol
Mexico	4/1/85	9/14/87	9/16/87	3/31/88
Morocco	2/7/86		1/7/88	
Netherlands	3/22/85	9/28/88	9/16/87	12/16/88
New Zealand	3/21/86	6/2/87	9/16/87	7/21/88
Nigeria	*	10/31/88	10/31/88	10/31/88
Norway	3/22/85	9/23/86	9/16/87	6/24/88
Panama	*	2/13/89	9/16/87	
Peru	3/22/85			
Philippines			9/14/88	
Portugal	*	10/17/88	9/16/87	10/17/88
Senegal			9/16/87	
Singapore	*	1/5/89	*	1/5/89
Spain	*	7/25/88	9/21/88	12/16/88
Sweden	3/22/85	11/26/86	9/16/87	6/29/88
Switzerland	3/22/85	12/17/87	9/16/87	12/28/88
Thailand			9/16/88	
Togo			9/16/87	
Uganda	*	6/23/88	9/15/88	9/15/88
Ukranian SSR	3/22/85	6/18/86	2/18/88	9/20/88
USSR	3/22/85	6/18/86	12/29/87	11/10/88
United Kingdom	5/20/85	5/15/87	9/16/87	12/16/88
USA	3/22/85	8/27/86	9/16/87	4/21/88
Venezuela	*	9/1/88	9/16/87	2/6/89
Total – without EEC	27	40	46	33

Entry into force for the Montreal Protocol requires that 20 nations ratify the Vienna Convention and 11 countries, representing two thirds of the world's CFC and halon consumption, ratify the Montreal Protocol.

*Ratification by accession (without signing)

Dates indicate when the instrument of ratification was deposited at the United Nations.

Source: Liz Cook (FoE US), Editor *Atmosphere*. Position as at 1 April 1989

APPENDIX 4

Renewable energy sources

The sun's energy maintains the temperature of the Earth, enabling plants and animals to thrive. Without this continual input of energy, surface temperatures would plummet and the planet become a barren and inhospitable wasteland.

1 Sunlight

The sun is a massive nuclear furnace radiating energy into space. One thousandth of a millionth of the sun's output of around 400,000,000,000,000,000,000,000,000 watts is intercepted by the earth. About 30 per cent of this energy is reflected back into space. The rest is absorbed by atmosphere, land, and oceans, or in evaporation, convection and precipitation of water.

2 Wind and waves

The energy in wind and waves is also solar energy. Heating the Earth's surface causes high and low air pressures and makes air move. The wind whips up the seas into waves. The energy that rustles leaves on trees – and occasionally blows them over – is solar energy. So too is the energy of ocean waves crashing on the shore and sending sea spray flying.

3 The tides

The rise and fall of the tides follows a regular and entirely predictable pattern. The tidal range is relatively small in mid-ocean, but a continental shelf, or the funnelling effect of bays, estuaries and straits that impede the tidal flow, amplifies its effects. Where the range is big enough, energy can be extracted using turbines mounted in a barrage.

The tides are due primarily to the gravitational pull from the moon. However, the position of the sun in the sky and the rotation of the Earth about its polar axis also play a part. The tides are highest when the earth, moon and sun are in line and lowest when they are pulling at right angles to one another.

4 Biomass

Material from living things, called 'biomass', stores solar energy. The sun's energy is used by trees and other green plants to manufacture simple sugars (from carbon dioxide in the air, and water in the soil) and convert these into more complex organic molecules like cellulose and lignin. When fuelwood is burnt for heat and light, the sun's energy is released. Likewise, food 'burnt' in our bodies is converted into heat and mechanical work. We too are powered by solar energy.

5 Running water

The sun evaporates water from oceans, lakes and rivers and carries the vapour up and over the land. Cooling leads to condensation – rain and snow to feed mountain streams and replenish the earth.

The sun in effect 'lifts' the water from the sea and deposits it on higher ground. The energy it acquires is exploited by hydro-electric installations. As the water falls back to sea it can turn a turbine and generate electricity.

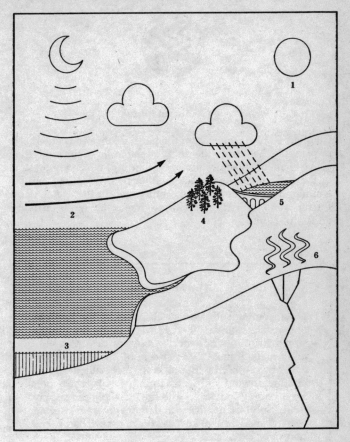

Fig. 18
Source: Friends of the Earth (1987)

6 Geothermal heat

Geothermal heat comes from the Earth's core and is enhanced in places by heat from the decay of naturally occurring radio-active elements (like uranium and thorium) which are present in the rock. It can be extracted either by drilling into natural aquifers and pumping out the geothermally heated water, or by forcing water under pressure through rock that has been deliberately fractured at depth.

The challenge

The incoming solar energy absorbed by the Earth in one year is equivalent to 15 to 20 times the energy stored in all of the world's reserves of recoverable fossil fuels. If just 0.005 per cent of this solar energy could be captured with fuel crops, specially designed buildings, wind and water turbines, solar collectors, wave energy converters and the like, it would supply more useful energy in a year than we get from burning coal, oil and gas. Unlike fossil fuels, renewable energy cannot be exhausted.

APPENDIX 5

CFCs – Ozone depletion potential and greenhouse strength

Chlorofluorocarbons (CFCs) are the chemicals primarily responsible for the depletion of the ozone layer, confirmed by the UK Ozone Trends Panel in March 1988. Ozone depletion poses a major threat to human health and the environment. CFCs are also recognised to be powerful 'greenhouse' gases, contributing to climate change.

They were invented in 1930 by Thomas Midgley of General Motors who also invented leaded petrol. CFCs have found a wide range of applications, particularly in the last fifteen years as a result of their essential properties:
- non-flammability
- low toxicity
- cheap and cost-effective
- extremely stable
- good thermal insulation properties
- waterproof insulation products in close cell structure.

The main uses of CFCs and the associated halon chemicals are in aerosols (30 per cent of world usage), refrigeration and air conditioning (30 per cent), foamed plastics (34 per cent), and a range of other uses including the cleaning of electronic circuit boards and fire extinguishers (6 per cent). These percentages will undoubtedly change rapidly over the next few years. Aerosol usage will drop to virtual zero within a few years if industry and countries fulfil the commitments they are currently making. Friends of the Earth *Alternatives to CFCs* (see Bibliography) is a useful guide to many of the main alternatives available for these chemicals.

CFC11 and 12 are the most common chemicals used at present. They are currently being replaced by so-called 'soft' CFCs, which either have a much lower ozone depletion potential or none at all. CFCs and halons are not only ozone depleters, they are also extremely powerful greenhouse gases, up to 24,000 times more effective in absorbing infra-red radiation than a single molecule of carbon dioxide. It is extremely important that a switch to 'soft' CFCs does not lead to a rapid growth in their use and an exacerbation of global warming. In the words of Kai Millyard of FOE Canada, this would be like 'jumping from the frying pan and into the oven'.

The following calculations on a range of CFCs and alternatives has been provided by Du Pont, who along with ICI, produce the bulk of these chemicals worldwide. A figure of 0.05 (eg, HCFC22) means that the chemical has only 5 per cent the ozone depletion potential of CFC11. A figure of 0.1 for greenhouse strength will still rank the chemical (eg, HFC134a) as having 1,000 times greater strength than carbon dioxide.

Substance	Chemical Formula	Ozone Depletion Potential	Greenhouse Strength
CFC12	CCl_2F_2	0.9	1.0
HCFC22	$CHClF_2$	0.05	0.07
HCFC123	$CHCl_2CF_3$	<0.15	<0.1
HCFC124	$CHClFCF_3$	<0.5	<0.1
HFC125	CHF_2CF_3	0	<0.2
HFC134a	CFC_3H_2F	0	<0.1
HCFC141b	CH_3CCl_2F	<0.05	<0.1
HCFC124b	CH_3CClF_2	<0.05	<0.2
HFC152a	CH_3CHF_2	0	<0.1

APPENDIX 6

Organisations

American Council for an Energy Efficient Economy (ACEEE)
1001 Connecticut Avenue, NW,
Washington DC, 20036,
USA.
Carry out research, policy initiatives and lobbying on all aspects of energy efficiency. Have lots of information on electrical and gas appliances and standards.
[Contact: Howard Geller (Director)]

Association for the Conservation of Energy (ACE)
9 Sherlock Mews,
London W1M 3RH,
United Kingdom.
tel: (01) 935 1495
Policy, research and lobbying organisation, promoting greater adoption of energy efficiency measures in the United Kingdom and the European Community. Produce a free newsletter and also run an energy and environment programme, concentrating on the greenhouse effect.
[Contact: Andrew Warren (Director), Stewart Boyle]

Australian Conservation Foundation

672 B Glenferrie Road,
Hawthorn 3122
Australia.
tel: (61) 3 819 2888
Research and campaigning organisation on a wide range of environmental issues. Very active on CFCs and greenhouse.
[Contact: Bill Hare]

Beijer Institute

The International Institute for Energy and Human Ecology,
The Royal Swedish Academy of Sciences,
Box 50005 S-10405 Stockholm,
Sweden.
tel: (08) 16 0490
Co-ordinate and advise many of the committees working at a policy level on climate change, as well as carrying out a great deal of energy and climate related research.
[Contact: Gordon T Goodman, Jill Jäger]

British Association of Nature Conservation (BANC)

Rectory Farm,
Stanton St. John,
Oxford OX9 1HF.
tel: (086735) 214.
Hold meetings and seminars on conservation issues and run many practical projects. Produce the journal *Ecos* and discussion papers.

Carbon Dioxide Information Analysis Centre (CDIAC)

Oak Ridge National Laboratory,
PO Box 2008,
Oak Ridge,
Tennessee 37831-6050,
USA.
tel: (615) 574 0390
Provide information, a newsletter and computer modelling packages relating to CO_2 research on request. Funded by the US Department of Energy.
[Contact: Tom Bowden, Laura Morris]

Centre for Alternative Technology
The Quarry,
Near Machynlleth,
Powys,
Wales.
tel: (0654) 2400
Practical demonstration of energy efficiency and renewable
energy. Large outdoor exhibition and bookshop. Also run
practical courses.
[Contact: Lesley Bradman (courses), Andy Rowland]

Centre for Applied Climatology and Environmental Studies
Department of Geography,
University of Munster,
D-4400 Munster,
Robert-Koch-Str 26,
Deutschland.
Research into reduction of greenhouse gases.
[Contact: Professor Wildfrid Bach]

The Centre for Our Common Future
Palais Wilson,
52, rue des Pâquis,
CH - 1201 Geneva,
Switzerland.
tel: (022) 32 71 17
The office set up after the work of the World Commission for
Environmental and Development officially ceased in 1987.
[Contact: Chip Lindler]

Climatic Research Unit (CRU)
School of Environmental Sciences,
University of East Anglia,
Norwich NR4 7TJ,
United Kingdom.
tel: (0603) 592091
One of the world's foremost climatic research centres.
[Contact: Tom Wigley (Director), Mick Kelly and Dick Warwick]

Department of the Environment UK (DoE)
2 Marsham Street,
London SW1P 3EB
United Kingdom.
tel: (01) 276 3000
UK's government lead department on the greenhouse effect.
[Contact: – Chief Scientist Dr D. Fisk (also the Director of the
Air, Noise and Wastes Directorate);
– Ozone – Fiona McConnell;
– Greenhouse Effect – Alan Apling, John Merliss.]

Energy Efficiency Office
Department of Energy,
Thames House South,
Millbank,
London SW1P 4QJ.
tel: (01) 211 3000
Co-ordinate government promotion of energy efficiency.
[Contact: Elliott Finer (Director General)]

Energy Probe
100 College Street,
Toronto,
M5G 1L5,
Canada.
tel: (416) 978 7014
Research on energy and climate change.
[Contact: Norman Rubin]

Environmental and Energy Study Institute
122 C Street NW,
Suite 700,
Washington, DC 20001,
USA.
tel: (202) 628 1400
Carry out policy orientated research which feeds into the US
Congress. Produce regular bulletins and reports on the political
status of climate and other issues.
[Contact: Jon Clark, Carol Werner]

Environmental Defense Fund

1. National Headquarters,
   257 Park Avenue South,
   New York, NY 10010,
   USA.
   tel: (212) 505 2100

2. Washington Office,
   1616 P Street, NW,
   Washington, DC 20036,
   USA.
   tel: (202) 387 3500

Conduct research and campaigning programme on global atmospheric issues.

[Contact: Michael Oppenheimer]   [Contact: Jo Goffman]

Environmental Liaison Centre (ELC)

PO Box 72461,
Nairobi,
Kenya.
tel: (Nairobi) 24770/340349
Co-ordinates non-governmental organisations in the developing world.
[Contact: Ann Heidenreich]

Environmental Protection Agency (EPA)

Division of Global Change,
Office of Air and Radiation,
ANR 445,
Washington, DC 20460,
USA.
An independent United States federal agency established in 1970, with responsibility for reducing pollution. Major research work on ozone depletion and climate change.
[Contact: James Titus (sea-level change), Michael Gibbs (CFCs), Bill Reilly (Administrator)]

Enquete Kommission

Deutscher Bundestag,
Bundeshaus,
5300 Bonn 1,
Deutschland.
Major Bundestag Inquiry Commission into ozone layer and climate change. Final Report due in late 1989/early 1990.
[Contact: Bernd Schmidbauer, Peter Henneke]

European Commission
Rue de la Loi 200,
1049 Brussels,
Belgium.
tel: 235 11 11
The permanent officials of the European Community, organised into a number of Directorates each reporting to a Commissioner, usually a politician appointed by the member states. Permanent civil servants staff the Directorates, each under a Director General; the main Directorates involved in climate change are:

DG XI   (Environment, Consumer Protection and Nuclear Safety); Commissioner: Mr Carlo Ripa di Meana. [Contact: Pierre Delogu]

DG XII   (Science, Research and Development). [Contact: Mr P. Bourdeau (Energy), Mr Fantechi (Climatology)]

DG XVII   (Energy); Commissioner: Antonio Cardosé Cunha. [Contact: Clive Jones, Peter Faross, and Derek Fee]

European Environmental Bureau
20 Rue de Luxembourg,
1040 Brussels,
Belgium.
tel: (2) 514 1250
Co-ordinate work by some of the European NGOs in the European Community, and provide an information service. [Contact: Annie Roncerel]

Friends of the Earth International (FoE)
Secretariat based at FoE UK offices:
Friends of the Earth,
26–28 Underwood Street,
London N1 7JQ,
United Kingdom.
tel: (01) 490 1555
Environmental pressure group campaigning on a wide range of issues including acid rain, nuclear power, CFCs and the greenhouse effect. The Secretariat is based in London and co-ordinates activities between the 35 Friends of the Earth groups.

[Contact: Nicole Mueller. UK Staff – Koy Thompson (Tropical Forests), Fiona Weir (CFCs), Simon Roberts (Energy); US – Liz Cook (CFCs and editor of *Atmosphere*, a quarterly newsletter) – tel: (202) 544 2600; CANADA – Kai Millyard (CFCs and Global Warming) – tel: (613) 230 3352]

Gaia Foundation
18 Well Walk,
Hampstead,
London NW3 1LD,
United Kingdom.
tel: (01) 435 5000
Work closely with James Lovelock and Jose Lutzenberger, particularly on the tropical rainforest issue.
[Contact: Liz Hoskens]

Green Alliance
60 Chandos Place,
London WC2N 4HG.
tel: (01) 836 0341
A cross-party environmental lobbying initiative for the UK.
[Contact: Julie Hill, Tom Burke (Director)]

Green Net
26–28 Underwood Street,
London N1 7JQ,
United Kingdom.
tel: (01) 490 1510
Global computer network, providing communication facilities to the environmental, peace and human rights movements. Has energy, environment and climate change 'conferences'.

Greenpeace International
Temple House,
25–26 High Street,
Lewes,
East Sussex BN2 7LU.
tel: (0273) 57878
Co-ordinates international operation of Greenpeace national groups. Campaigns on toxic waste, protection of cetaceans,

nuclear power, air pollution (including CFCs and global warming), and marine ecology.

[Contact: Andy Kerr, Andrew Stirling; UK – Steve Elsworth tel (01) 354 5100]

International Atomic Energy Agency (IAEA)

Wagramerstrasse 5,
PO Box 100,
A-1400 Vienna,
Austria.
tel: (1) 23 600

Formed in 1957, it now has 113 Member States. Its objectives are to accelerate and enlarge the global contribution of atomic energy and to safeguard fissile materials. Currently carrying out work on the greenhouse effect.

International Council of Scientific Unions (ICSU)

51, Boulevard de Montmorency,
PARIS, F - 75016,
France.

Through the Scientific Committee on Problems of the Environment (SCOPE), it tries to assemble, review and assess the information on man-made environmental changes and the impacts of these changes, to evaluate the methodologies of measurement of environmental change and to provide information on current research. SCOPE 29 (see reading list) is one of the best source documents on global warming and climatic change.

International Energy Agency (IEA)

a) Coal Research
14–15 Lower Grosvenor Place,
London SW1W 0EX,
United Kingdom.
tel: (01) 828 9508
[Contact: Irene Smith]

b) PARIS
Château de la Nuette,
2, Rue André Pascal,
Paris, FRANCE.
tel: (Paris) 524 8200
[Contact: Bill Long, Sergio Garriba]

Carry out research and policy discussions on energy efficiency, coal use, greenhouse emissions.

International Institute for Applied Systems Analysis (IIASA)
A-2361,
Laxenburg,
Austria.
tel: (36) 715 216.
Policy institute, working on impact assessments of global warming, especially agricultural related issues. Also produce energy scenarios.
[UK contact: Martin Parry,
                Atmospheric Impact Research Group,
                School of Geography,
                University of Birmingham,
                B15 2TT.
                tel: (021) 414 5548.]

International Institute for Energy Conservation
420 C Street, NE,
Washington, DC 20002,
USA.
tel: (202) 546 3388
Initiate practical ways of encouraging energy efficiency in the developing world. Has good contacts in China and Indonesia.
[Contact: Deborah Bleviss (Director)]

International Institute for Environmental Development (IIED)
3 Endsleigh Street,
London WC1H 0DD,
United Kingdom.
tel: (01) 388 2117
IIED is an organisation which is heavily involved in 'sustainable development' issues and practical projects in the Third World. It produces a regular newsletter and publishes books through *Earthscan*.
[Contact: Lloyd Timberlake, Richard Sandbrook (Director)]

International Organisation of Consumer Unions (IOCU)
PO Box 1045,
10830 Penang,
Malaysia.
tel: (011) 60 04 20391
Major NGO campaigning network in South-East Asia. Areas of work include tropical rainforests and development.
[Contact: Martin Abraham, Chee Yoke Ling]

Inter-governmental Panel on Climate Change (IPCC)
c/o WMO (see below).
tel: (22) 34 6400
[Contact: Dr N. Sundararaman]

Men of Trees
HQ,
Crawley Down,
Crawley,
Sussex.
tel: (0342) 712536.
Support and provide information on the protection and development of tree projects.

Meteorological Office
London Road,
Bracknell,
Berkshire RG12 2SZ.
tel: (0344) 420242
Runs one of the main GCMs in the world. John Houghton is co-ordinating one of the three IPCC committees on the science of global warming.
[Contact: John Houghton, Geoff Jenkins]

Milton Keynes Energy Park
Milton Keynes Development Corporation,
Saxon Court,
502 Avebury Boulevard,
Central Milton Keynes MK9 3HS.
tel: (0908) 692692.
Encourage the construction of houses using highly energy
efficient and solar building designs and techniques, and have
developed an Energy Conservation Index for houses.
[Contact: Stephen Fuller]

National Centre for Atmospheric Research (NCAR)
PO Box 3000,
Boulder, CO 80307,
Colorado, USA.
tel: (303) 497 1000
Carry out computer modelling and run one of the four main
GCMs in the world. Have recently produced new research on
methane.
[Contact: Stephen Schneider, Ralph Cicerone]

Natural Resources Defense Council
1350 New York Avenue NW,
Washington, DC 20005,
USA.
tel: (202) 783 7800
Research and policy initiatives on ozone depletion and climate
change.
[Contact: Jacob Scherr, Richard E. Ayres, David Doniger
(CFCs)]

NASA Goddard Space Flight Centre
Institute for Space Studies
2880 Broadway,
New York, NY 10025,
USA.
Carry out computer modelling and run one of the four main
GCMs in the world. Also at the forefront of work on ozone
depletion.
[Contact: Jim Hansen, Bob Watson]

Neighbourhood Energy Action (NEA)

Energy Projects Office,
2nd Floor,
Sunlight Chambers,
2–4 Bigg Market,
Newcastle-upon-Tyne NE1 1VW
tel: (0632) 615677

Run energy conservation schemes around United Kingdom and can provide advice to groups interested in setting up new projects.
[Contact: Andrea Cook]

Nuclear Information and Resource Service (NIRS)

1424, 16th Street,
NW Suite 601,
Washington, DC 20036,
USA.
tel: (202) 328 0002

Provides circular and information on all aspects of nuclear industry, concentrating on nuclear waste issue.
[Contact: Diane D'Arrigo, Jim Riccio]

Office of Technological Assessment (OTA)

Congress of the United States,
Washington, DC 20510,
USA.
tel: (202) 228 6845

Advises Congress on environmental and technological issues. Researches into climate change.
[Contact: Dr Rosina Bierbaum, Nick Sundt, Oceans and Environment Programme.]

Overseas Development Administration (ODA)

Natural Resources Unit,
Eland House,
Stag Place,
London SW1E 5DH.
tel: (01) 273 3000

British government body with responsibility for development abroad.

Oxfam
27 Banbury Road,
Oxford OX2 7DZ,
United Kingdom.
tel: (0865) 56777
Campaigns on development work with a focus on famine relief and aid to the Third World.

Rocky Mountain Institute
1739 Snowmass Creek Road,
Snowmass,
Colorado 81654-9199,
USA.
tel: (303) 927 3128
Policy research centre for global and national energy issues. Runs *Competitek* to advise on efficient end-use electricity.
[Contact: Amory Lovins, Ted Flannigan]

Survival International
310 Edgware Road,
London W2 1DY.
tel: (01) 725 5535
International human rights organisation, which works for the rights of threatened tribal peoples, including those in tropical forests.

TATA Energy Research Institute
7 Jor Bagh,
New Delhi, 110003,
India.
tel: (61) 9205 615032
Research and policy analysis, including energy efficiency
[Contact: Dr Ravendra Pachauri]

Tree Council
Agricultural House,
Knightsbridge,
London SW1X 7NJ.
tel: (01) 235 8854.
Aims to promote planting. Supports practical projects and provides information.

United Nations Environment Programme (UNEP)

1. UNEP,
   PO Box 47074,
   Nairobi,
   Kenya.
   tel: (Nairobi) 333930

2. UK UNEP,
   c/o David Hall,
   IIED (see above),

[Contact: Peter Usher, Mostofa Tolba]

Created after the 1972 Stockholm UN Conference on the Human Environment to provide the focus for world-wide discussion and action on urgent environmental threats and international monitoring.

War on Want

37–39 Great Guilford Street,
London SE1 0ES,
United Kingdom.
tel: (01) 620 1111

Supports grass roots movements for change in developing countries, and campaigns in Britain on issues of world poverty.
[Contact: Helen O'Connell]

Women's Environmental Network (WEN)

287 City Road,
London N1.
tel: (01) 490 2511/251 0149.

Campaigns on wide range of environmental issues as they affect women.
[Contact: Bernadette Vallely, Alison Costello]

Woods Hole Research Centre

PO Box 296,
Woods Hole, MA 02543,
USA.
tel: (617) 540 9900

Carry out legal and policy research on climate change as well as work on the science of the greenhouse effect.
[Contact: Dr Kiliparti Ramakrishna, George Woodwell (Director)]

World Meteorological Organisation (WMO)
Case Postale No. 5,
1211 Geneva, 20,
Switzerland.
tel: (22) 34 6400
[Contact: Dr N. Sundaraman, Dr Milton Obasi (Director-General)]

World Resources Institute (WRI)
1735 New York Avenue, NW,
Washington, DC 20006,
USA.
tel: (202) 638 6300
Policy research centre, created in 1982, to help policy-makers on issues such as sustainable development and natural resources. Current work includes global energy futures and modelling on climate change.
[Contact: Rafe Pomerance (policy and lobbying), Irving Mintzer (research and policy)]

World Wide Fund for Nature (WWF)
Panda House,
Weyside Park,
Guildford GU7 IX7,
United Kingdom.
tel: (04834) 26444
An international organisation concerned with a wide range of issues ranging from endangered species protection to acid rain and tropical rainforests.
[Contact: Janet Barber, Tessa Robertson]
[Also: WWF International; contact Adam Markham tel: Geneva 2264 71 81]

GLOSSARY AND ABBREVIATIONS

analog (weather) A large scale weather pattern of the past which is similar to a current situation in its essential characteristics

albedo The reflectiveness of a surface expressed as the ratio of the light reflected to the total light received. Snow has a high albedo, dark soils have a low albedo

atmosphere The envelope of air surrounding the Earth and bound to it by the Earth's gravitational attraction

biomass The total biological matter, or stored energy content of living organisms, existing in a given specified volume or area. Often referred to when considering the potential production of available energy for food or fuel

Boreal Period The period prior to 5000 BC (ie, preceding the Atlantic Period), when the climate was of the Continental type, with summers much warmer than at present

carbon budget The balance of the exchanges (incomes compared to losses) of carbon, either between the carbon reservoirs, or one specific route (eg atmosphere to the biosphere) of the carbon cycle

carbon sink The mass of carbon absorbed or tied up in the carbon cycle

climate The statistical collection and representation of the weather conditions for a specified area during a specified time interval, usually decades or more. The properties which characterise the climate include thermal (surface air temperatures, water, land, ice), kinetic (wind and ocean currents, together with humidity, cloudiness) and static (pressure and density of atmosphere)

condensation In meteorological usage the term is applied to the production of water vapour by the cooling of saturated air. This may occur when warm air rises to higher, colder levels (eg by rising over mountains) or when the warm air in a room touches a cold window and water droplets form

deoxyribonucleic acid (DNA) A large coiled biological molecule which contains genetic material. It is the basic building block of animal life and determines our sex, shape, colour and other characteristics. Damage to the DNA can lead to cancers, malformations and other problems

El Niño An irregular variation of ocean current that flows off the west coast of South America from January to March, carrying warm, nutrient-poor water of low salinity, in a southerly direction

Environmental Impact Assessment (EIA) The European extension (Directive 85/37/EEC) of the USA's environment impact statement. It requires an analysis and judgment of the effects upon the environment (including climate change), both temporary and permanent, of a significant development or project. It must also consider the social consequences and alternative actions

evaporation The transition from liquid to gas

fast breeder reactor (FBR) A nuclear reactor using plutonium as a fuel that theoretically produces more fuel than it consumes. Plutonium releases fast neutrons (hence the name Fast Breeder Reactor) which bombard uranium obtained from spent fuel from conventional reactors, to produce more plutonium. No moderator is required and liquid sodium is usually used as a coolant

fissile material Material which is readily capable of undergoing fission when struck by a neutron, eg Uranium U-235

general circulation models (GCMs) Modelling conducted on computers to enable analysis and prediction of carbon flows in the carbon cycle, future climatic patterns and characteristics, and the impacts of climatic change

greenhouse effect Because the Earth's atmosphere is transparent to much solar radiation, this heats up the Earth's surface and infra-red radiation is emitted. This is partially trapped in the atmosphere by carbon dioxide, water vapour and trace gases. The effect on Venus is far more pronounced due to much higher levels of carbon dioxide

hectare (ha) Unit of area: $1ha = 3.47$ acres, $100ha = 1km^2$

hydrological cycle The process of evaporation, vertical and horizontal transport of vapour, condensation, precipitation and the flow of water from continents to oceans

International Council of Scientific Unions (ICSU) (see Organisations)

infra-red radiation Electromagnetic radiation lying in the wavelength interval from 0.7 to 1000 Im. Its lower limit is bounded by visible radiation and its upper limit by microwave radiation

megawatt (MW) 10^6 watts. A large coal or nuclear station has an electrical generating capacity of around 1000MW

mtce Million tonnes coal equivalent

photosynthesis The process by which green plants make carbohydrates from water and carbon dioxide, using the sun's energy

plutonium (Pu) An element not naturally occurring, but produced during nuclear reactions. It forms several isotopes – ^{238}Pu, ^{239}Pu, ^{240}Pu, ^{241}Pu and ^{242}Pu – each with varying properties and uses, including nuclear weapons and as fuel for FBRs

precipitation A term used in meteorology for all forms of atmospheric moisture such as rain, hail, snow, sleet and dew

respiration A biochemical process by which living organisms take up oxygen from the environment and, in the case of plants, consume some of the organic matter produced by photosynthesis during daylight hours. In the case of animals, both carbon dioxide and heat are released during respiration

stratosphere The layer in the Earth's atmosphere above the troposphere, which is stable and has very low humidity

tropopause The transition between the troposphere and the stratosphere

troposphere The inner layer of the Earth's atmosphere, going up to 15km at the equator and 8km in the polar regions, in which nearly all clouds form and the weather is determined

ultraviolet (UV) radiation Electromagnetic radiation with wavelengths less than the visible spectrum but longer than X-rays. It is non-ionising but can be damaging to tissues. Much of the solar radiation in this band is filtered out by the ozone in the upper atmosphere

UNEP United Nations Environment Programme (see Organisations)

uranium (U) Naturally occurring radioactive metallic element, the basis of all nuclear power generation. It is widely dispersed, but in low concentrations in most ores, and exists in three different natural isotopes

WMO World Meteorological Organisation (see Organisations)

BIBLIOGRAPHY

This brief reading list should allow a more detailed insight into the global warming issue. Those references marked with an asterisk (*) are essential reading. *New Scientist*, *Nature*, *Energy Policy*, *Climatic Change* and *Science* are the best general periodicals for keeping up to date with the subject.

1. General Reading on Global Warming and the Environment

Gaia, A New Look At Life On Earth, J. Lovelock, Oxford University Press, 1979

*The Gaia Atlas of Planet Management, Norman Myers, (Ed), Pan Books, 1986

The Ages of Gaia, J. Lovelock, Oxford University Press, 1988

Thinking Like a Mountain: towards a a Council of All Beings, John Seed, Joanna Macy, Pat Fleming, Arne Naess, 1988, Heretic Books

The Co-evolution of Climate and Life, S. Schneider, R. Londer, Sierra Club Books, 1984

The Earth Report, Edward Goldsmith and Nicholas Hildyard, (Eds), Mitchell Beazley, 1988

Earthscope, Chris Pellant, (Ed), Salamander Books, 1987

Ecotopia, and *Ectopia Revisited*, E. Callenbach, Pluto Press, 1978

*Effects of Changes in Stratospheric Ozone and Global Climate, (4 Volumes), J. G. Titus, (Ed), Environmental Protection Agency/United Nations Environment Programme, Washington DC, 1986

*The Greenhouse Effect, Climatic Change and Ecosystems, B. Bolin and others, (Eds), (SCOPE 29), Wiley, 1986

The Greenpeace Book of Antarctica, John May, Dorling Kindersley, 1988.

*State of the World, Worldwatch Institute, Washington, 1987, 1988 and 1989

World Resources, 1986, 1987 & 1988, World Resources Institute and IIED

2. Weather
Weather, Climate and Human Affairs, H. H. Lamb, Routledge and Kegan, 1988
Weather Systems, L. F. Musk, Cambridge Educational, 1988

3. Ozone/CFCs
Alternatives to CFCs, Friends of the Earth, 1989
As Safe as Houses: CFCs in Buildings, Friends of the Earth, 1989
Future Concentrations of Stratospheric Chlorine and Bromine, J. Hoffman, US EPA (EPA 400/1-88/005), July 1988
The Hole in the Sky, John Gribbin, Corgi, 1988
Ozone Depletion: Health and Environmental Consequences, R. R-Jones & T. Wigley, (Eds), John Wiley, (due out Autumn 1989)
The Ozone Layer, UNEP/GEMS Environment Library, No 2, 1988
Protecting Life on Earth: Steps to Save the Ozone Layer, Cynthia Pollock Shea, Worldwatch Paper No.87, Worldwatch Institute, 1988
Stratospheric Ozone, 1st and 2nd Reports of the UK Stratospheric Ozone Review Group, HMSO, 1987–88

4. Nuclear Power
Energy/War: Breaking the Links, a Prescription for Non-proliferation, A. Lovins & H. Lovins, Harper Colophon, 1981
The End of the Nuclear Dream: Nuclear R + D and the future of the UKAEA, M. Flood, Friends of the Earth, 1988
The Future of Nuclear Power, G. Greenhalgh, Graham and Trotman, 1988
Going Critical: An Unofficial History of British Nuclear Power, W. Patterson, Paladin, 1986
Nuclear Power, W. Patterson, Pelican, 1985
Power at Any Price: France and Nuclear Power, P. Davies, Friends of the Earth, 1986
Radiation and Health, R. R-Jones and R. Southwood, (Eds), Wiley, 1987
The Realities of Nuclear Power: International economic and regulatory experience, S. Thomas, Cambridge University Press, 1988

5. General Energy

(i) Efficient use of fossil fuels

Advanced Coal Use Technology, W. Patterson, Financial Times Business Information, 1987

The Clean Use of Coal: A technology review, IEA/OECD, 1985

(ii) Energy conservation

Combined Heat and Power, Energy Paper No 35, United Kingdom Department of Energy, London, 1979

Energy Conservation in IEA countries, OECD, 1987

Energy Conservation in Japanese Industry, R. Dore, British Institutes' Joint Energy Policy, Energy Paper No 3, 1983

Energy Efficiency: A New Agenda, W. Chandler, H. Geller, M. Ledbetter, The American Council for an Energy-Efficient Economy, July 1988

Energy Efficient Buildings, A. H. Rosenfeld and D. Hafemeister, Scientific American, 258 (4), April 1988

*Energy for a Sustainable World, J. Goldemberg, R. Williams, A. Reddy, T. Johannsen, World Resources Institute, Washington DC, 1987 (The full report is printed by John Wiley, 1988)

*Energy Productivity: The Key to Environmental Protection and Economic Progress, William U. Chandler, Worldwatch Paper 63, January 1985

*Global Forces towards greater Efficiency, R. Malpas, 1989 (available from BP-UK)

Lessons from America, from Association for the Conservation of Energy, 1985–87

More Efficient Use of Electricity – the Megawatt resource, D. Olivier, Paper presented in seminar on Energy and Climatic Change – What can Europe Do?, Brussels, June 1988

*The New Oil Crisis and Fuel Economy Technologies: Preparing the Light Transportation Industry for the 1990s, D. Lynn Bleviss, Quorum Books, 1988

*Rethinking the Role of the Automobile, M. Renner, Worldwatch Paper 84, June 1988

Transport and the Environment, OECD, 1988

(iii) Renewables

Alternative Energy Sources, House of Lords Committee on European Communities, 16th Report, 1987–88

*Energy Without End: the Potential for Renewable Energy, M. Flood, Friends of the Earth, London, 1987

Renewable Energy Sources for the 21st Century, R. Hill, P. O'Keefe and N. M. Pearsall, (Eds), Adam Hilger, 1988
Renewable Energy: Today's Contribution, Tomorrow's Promise, C. Pollock Shea, Worldwatch Paper No 81, January 1988

(iv) Reductions in Carbon Dioxide through efficiency
Electricity for Life?: Choices for the Environment, J. Skea, Friends of the Earth/Council for the Protection of Rural England, 1988
Greenhouse Warming, Comparative Analysis of Two Abatement Strategies, B. Keepin and G. Kats, *Energy Policy*, December 1988
A Matter of Degrees, Irving Mintzer, World Resources Institute, 1987
Strategies to reduce greenhouse emissions in the UK, I. Brown & S. T. Boyle, Association for the Conservation of Energy (ACE), 1989

6. Environment and Development
Africa in Crisis, Lloyd Timberlake, Earthscan, 1986
Beyond the Woodfuel Crisis, Leach and Mearns, Earthscan, 1989
Our Common Future, World Commission on Environment and Development, Oxford University Press, 1987
The UK Government's Response to the Brundtland Report, Department of the Environment, 1988
Britain and the Brundtland Report, (Variety of Development and Environmental groups from IIED), 1988
Brundtland in the Balance: a critique of the UK Government's response to the WCED, various groups, IIED, 1989
Energy and Growth: A Comparison of 13 Industrial and Developing Countries, G. Leach, L. Jarass, G. Obermair, L. Hoffmann, Butterworth, 1980
The Fuelwood Trap, Munslow and others, Earthscan, 1986
The Greening of Africa, P. Harrison, Paladin, 1987
The Greening of Aid, C. Conroy and M. Litvinoff, (Eds), Earthscan, 1988
Women and Environment in the Third World, I. Dankelman and J. Davidson, Earthscan, 1988

7. Greenhouse Gases
Greenhouse Gases, UNEP/GEMS Environment Library, Number 1, 1987
Altering the Earth's Chemistry: Assessing the Risks, S. Postel, Worldwatch Paper 71, July 1986
The Changing Atmosphere, F. S. Rowland and I. S. A. Isaksen, (Eds), J. Wiley, 1988

CO₂ and Climate Change, Irene Smith, International Energy Agency, 1988

The Effect of CO₂ Emissions from Coal Fired Power Plants, K. M. Sullivan, International Coal Development Institute, 1988

Review of the greenhouse theory, V. Ramanathan, Science, vol. 240, p.293–299, 1988

State of the Art Report on the Carbon Dioxide Research Programme, US Department of Energy, 1985

8. Impacts

Glaciers, Ice Sheets and Sea Level: Effect of a Carbon Dioxide-Induced Climatic Change, US Department of Energy, Reidel, 1984

The Impact of Climatic Variations on Agriculture, Parry, Carter and Konijn, (Eds), IIASA, 1988

Possible impacts of Climate Change on the Natural Environment in the United Kingdom, United Kingdom Department of the Environment, London, 1988

The potential impacts of global climate change in the United States, Report to Congress, Smith and Tirpol, EPA, 1988

Scales of Climatic Impacts, W. Clark, Climatic Change, Vol. 7, p. 5–27, 1985

The Sensitivity of Natural Ecosystems and Agriculture to Climatic Change, W. R. Emmanuel and others, Edited by M. L. Parry, International Institute for Applied Systems Analysis (IIASA), Laxenburg, 1985

Waterlogged Wealth, Edward Maltby, Earthscan, 1986

See also SCOPE 29 (in General Reading)

9. Political/Policy

Air Pollution, House of Commons Environment Committee 1987–88, First Report, HMSO

Blueprint for the Environment: An Agenda for the new President, America's Environmental Community, 1988. (Available from NRDC and others)

The Greenhouse Effect: Issues for Policymakers, D. Everest, Policy Studies Institute/Royal Institute of International Affairs, 1988

Social Costs of Energy Consumption, O. Hohmeyer, Springer-Verlag, Berlin, 1988

Solutions to Global Warming: Questions and Answers, Stewart Boyle, ACE, 1989

10. Action

The Good Wood Guide, Friends of the Earth, 1988
*The Green Consumer Guide, J. Elkington and J. Hailes, Victor Gollancz, 1988
Home Ecology, K. Christensen, Arlington Press, 1989
See also Gaia Atlas, FoE Ozone Material and *The Earth Report*

11. Rainforests/Afforestation

*Biodiversity, E. O. Wilson, (Ed), National Academic Press, 1988
*The Prospect of Solving the CO_2 problem through Global Reforestation, G. Marland, US DoE (DoE/NBB0082), February 1988
*The Rainforest, C. Caufield, Heinemann, 1985
Save the Forests: Save the Planet, A Plan For Action, The Ecologist: Volume 17 No 4/5, 1987

INDEX

HENRY HAMMON AND STUART PARROTT

MAYDAY AT CHERNOBYL

On April 26th, 1986 at 1.23 am, an accident occurred that changed nuclear history.

MAYDAY AT CHERNOBYL tells the story of the world's worst nuclear disaster – why and how it happened – using Soviet and Western sources.

*Did the Soviet Union try to hide the truth about Chernobyl?

*Did Soviet nuclear scientists know before Chernobyl that the reactor could be dangerous?

*How many people will die of cancer because of the Chernobyl disaster?

*Can the environment recover?

MAYDAY AT CHERNOBYL is the first book to put the Chernobyl accident into context, showing how it will affect one of the world's superpowers and how both East and West used the accident for their own political ends.

Post·A·Book

A Royal Mail service in association with the Book Marketing Council & The Booksellers Association.

Post·A·Book is a Post Office trademark

JUDITH COOK

WHOSE HEALTH IS IT ANYWAY?
The Consumer and the National Health Service

'The National Health Service is safe in our hands'
Mrs Thatcher

THE GOVERNMENT VERSION:
More money spent, more hospitals, more patients treated – and certainly more statistics . . .

EVERYDAY EXPERIENCE:
Wards closed, operations cancelled, nurses in despair, endless waits and long waiting lists . . .

Once we had a health service to be proud of. Now it limps from crisis to crisis. So what *has* gone wrong?

Judith Cook, an investigative journalist long involved with Community Health Councils, suggests the time has come for some radical changes. We need a health service more concerned with preventative medicine and with better health education – rather than occasional outbursts of Edwina Currie-like lecturing – where the consumer – the ordinary person in the street – actually plays a part in the decision-making process.

WHOSE HEALTH IS IT ANYWAY? is for everyone who wants not just a patched-up NHS but a *better* NHS.

HODDER AND STOUGHTON PAPERBACKS

BEATA BISHOP

A TIME TO HEAL

In 1981, suffering from malignant melanoma, Beata Bishop was given six months to live. Today, after eight years of the unorthodox and controversial Gerson Therapy, she is fit and well and completely free of cancer.

The story of her triumph over such a virulent form of cancer using a diet-based therapy is both fascinating and alarming. Fascinating because of her remarkable recovery; alarming because of the questions it raises about our inability to deal with the terrifying spread of cancer in the Western world.

This book is an eloquent appeal to us to look seriously at a therapy which offers a genuine cure for cancer. It is also a profoundly moving story of quite extraordinary courage and determination.

HODDER AND STOUGHTON PAPERBACKS

ANDREW TYLER

STREET DRUGS

Prescribed, trafficked or stolen, drugs are swallowed, smoked, sniffed and injected at all levels of society. Patterns of use and abuse change and change again. Yet accurate information and level-headed advice are hard to come by.

Instead we are bombarded by a succession of panic headlines as scare follows scare. Even medical opinion reverses: yesterday's wonder prescription is today's killer. Fact and fiction are muddled. The law and government policies either limp along behind reality or become sidetracked by moral crusades. History is ignored and lessons never learned.

STREET DRUGS is a much-needed guide to the whole range of drugs – legal and illegal – in use today. Effects and side-effects, trade names and street names, history and geography, methods and fashions, benefits and dangers, are all clearly described.

STREET DRUGS is for drug workers, drug users, teachers, parents, for everyone who needs or wants to know about drugs and drug taking in the Eighties.

HODDER AND STOUGHTON PAPERBACKS

STEPHEN FULDER

THE HANDBOOK OF COMPLEMENTARY MEDICINE

THE HANDBOOK OF COMPLEMENTARY MEDICINE is already accepted as the most complete and authoritative guide to alternative medicine.

This new and completely revised edition includes all the latest developments in a very fast-growing area.

In a unique survey, all aspects of complementary medicine are covered, from the scientific to the social and the legal. Each therapy is described in detail, including its background and current practice.

Also included is an up-to-date list of organisations providing both treatment and study facilities.

'Exactly what is needed at the present time'
Sir James Watt,
Past President of the Royal Society of Medicine

'Contains the answers to the questions we have all asked, plus a lot more we hadn't thought of asking'
Health Now

'I suggest sceptics of complementary medicine read this book'
Nursing Times

HODDER AND STOUGHTON PAPERBACKS

MORE TITLES AVAILABLE FROM
HODDER AND STOUGHTON PAPERBACKS

☐	40858 2	HENRY HAMMON AND STUART PARROTT Mayday At Chernobyl	£2.95
☐	43109 6	JUDITH COOK Whose Health Is It Anyway?	£3.50
☐	50085 3	BEATA BISHOP A Time To Heal	£3.50
☐	42273 9	ANDREW TYLER Street Drugs	£4.99
☐	49484 0	STEPHEN FULDER The Handbook of Complementary Medicine	£5.99

All these books are available at your local bookshop or news-agent, or can be ordered direct from the publisher. Just tick the titles you want and fill in the form below.

Prices and availability subject to change without notice.

Hodder and Stoughton Paperbacks, P.O. Box 11, Falmouth, Cornwall.

Please send cheque or postal order, and allow the following for postage and packing:

U.K. – 55p for one book, plus 22p for the second book, and 14p for each additional book ordered up to a £1.75 maximum.

B.F.P.O. and EIRE – 55p for the first book, plus 22p for the second book, and 14p per copy for the next 7 books, 8p per book thereafter.

OTHER OVERSEAS CUSTOMERS – £1.00 for the first book, plus 25p per copy for each additional book.

Name ....

Address ....

....